# Phantom in Combat

# PHANTOM

# IN COMBAT
## WALTER J. BOYNE

Schiffer Military/Aviation History
Atglen, PA

Copyright © 1994 by Walter J. Boyne.
Library of Congress Catalog Number: 84-52013

All rights reserved. No part of this work may be reproduced or used in any forms or by any means – graphic, electronic or mechanical, including photocopying or information storage and retrieval systems – without written permission from the copyright holder.

Printed in the United States of America.
ISBN: 0-88740-599-1

This book was previously published by the Smithsonian Institution Press, & Jane's Publishing Co., 1985.

We are interested in hearing from authors with book ideas on related topics.

Published by Schiffer Publishing Ltd.
77 Lower Valley Road
Atglen, PA 19310
Please write for a free catalog.
This book may be purchased from the publisher.
Please include $2.95 postage.
Try your bookstore first.

## CONTENTS
FOREWORD BY LT.COL. STEVE RITCHIE 6
ACKNOWLEDGEMENTS 7
INTRODUCTION 9
CHAPTER 1: A TASTE OF COMBAT 10
CHAPTER 2: GHOSTLY LINEAGE 22
CHAPTER 3: DESIGNING THE PHANTOM 31
CHAPTER 4: VIETNAM: THE CONTEXT OF COMBAT 45
CHAPTER 5: ACES AND ISSUES 71
CHAPTER 6: NAVAL AND MARINE PHANTOMS IN VIETNAM 81
CHAPTER 7: FIGHTING THE PHANTOM 106
CHAPTER 8: PHANTOM IN THE MIDDLE EAST 127
CHAPTER 9: A PEACETIME WINNER TOO 145
APPENDIXES 160

# Foreword
by Lt-Col Steve Ritchie

With unmatched style, feeling and insight Walter J. Boyne has tracked the F-4 Phantom II from its conception and birth, through an exciting development phase and a tremendously successful period of combat, to its present state as a fully mature weapons system and one of the world's leading fighter aircraft. In doing so he has described as no others have the senseless limitations placed on those who operated this fighting machine during its greatest test, the Vietnam War. Very few have had the guts to "tell it like it was" and place the blame for the "no win" war where it belongs – with the politicians.

Thousands of Americans, plus many allies, did what they had to under terribly difficult circumstances. And mostly they did a magnificent job. Thanks to the professionalism, determination, leadership and quality of the airmen mentioned in this book and many, many others, the F-4 prevailed under fire, and the success of the Phantom in combat was one of the few bright spots of an otherwise dismal period in American history. Inevitably, the war revealed some glaring shortcomings in the aircraft's design which adversely affected its operational performance. For instance, the single most important piece of equipment in air combat, the UHF radio, is located under the back seat. A radio change took two hours, entailed removal of the entire rear seat and almost always resulted in the cancellation of the sortie. The cockpit arrangement – particularly the positioning of the master arm and several other vital switches – is another example of an original design flaw. But these and other problems have all been solved over the years.

The question of two cockpits versus one is covered in some detail in this book. Most of us who flew the Phantom in combat, under a wide variety of circumstances and across a spectrum of missions and conditions, came to appreciate greatly the type's many advantages, and quickly learned to allow for its deficiencies. The Phantom was designed to be operated from two cockpits. There was equipment in the back that was not in the front, and a second crew member was required. Assuming that conditions of crew qualification, compatibility and co-ordination could be consistently met, the "guy in back" (or "GIB" as back-seaters were usually known) was a definite asset, particularly for special missions. I helped to develop the Fast FAC programme with F-4s based at Da Nang in 1968, and I can vouch for the fact that dual-cockpit capability proved invaluable in this highly successful operation.

However, what was a plus for some was a minus for others. Air Force policy during the war meant that no-one would be required to serve twice until everyone had been once. As a result, many who had never flown fighters or even knew the tactical mission, and many who had not flown for years, were suddenly rushed through five to six months of combat crew training and sent to South-east Asia. Quite often, on account of their rank, these men found themselves in combat leadership roles for which they were unqualified. This kind of mismatch had some dramatic and tragic results, particularly as the "no win" war began to drag on. The statement by an F-4 aircraft commander in Vietnam – "... I just sat up front and squeezed the trigger" – would never have been made by an aviator competent to operate a single-seat, high-performance, modern fighter in a highly complex combat arena. The current family of fighters and advanced tactical aircraft reflect changes which came as a result of this situation.

As Colonel Boyne points out, MacAir learned rapidly, listened to those who operated and came to know the aircraft, and was anxious to continue to improve its performance and reliability. In fact, many of the great advances built into the F-15 are the result of lessons learned in the F-4.

All in all, the Phantom proved itself in what was probably the most sophisticated defensive environment in the history of air combat and under the most restrictive operating constraints ever known. The F-4 clearly emerged as the "Queen of Battle" in South-east Asia, and along with so many others I am very proud to have crossed the "Red River" in the Phantom II.

# Acknowledgements

This book is written mainly from the point of view of the pilots who flew the Phantom in combat, yet it was made possible only by the willing co-operation of hundreds of people. Most of them were far removed from the action, yet they had indispensable information and insight, and were able to supply otherwise unavailable photos. I am very grateful to everyone who helped, and who endured my importunate requests for more information. I hope that anyone I have omitted will accept my apologies and my gratitude.

Let me thank first of all the many wonderful people of the McDonnell Douglas Aircraft Corporation who turned files and minds open to me, and then went through a laborious review of the manuscript. Special thanks must go to the entertaining and vastly knowledgeable Robert "Beaver" Blake, who exceeded his already international reputation for helpfulness, and without whom the book could not have been completed. The United States Air Force was as always extremely helpful, and I am grateful to my friend Brig-Gen Richard Abel, USAF Director of Public Affairs, for his special assistance. The US Navy and Marine Corps were equally obliging, as was the General Electric Corporation, where Len Dalquest and Ralph Wheeler combined to provide information and photos.

With regard to information on the Middle East, I particularly wish to thank the Honorable Richard Helms; Brig-Gen Menachem Eini, former Israeli air attaché in Washington; and Bruce Klein and Seffy Bodansky, who translated documents for me and provided insights into the psychology of the Middle Eastern conflicts.

Combat and ground crews provided stories, tapes and photos. They include, on the Air Force side, Lt-Col Steve Ritchie, who became the first USAF ace of the Vietnam War and kindly wrote the foreword to this book; Col Chuck DeBellevue, who scored six aerial victories in the Phantom; Col Albert Piccirillo, who combined tremendous combat experience with real photographic skill; Maj William Vasser; Dr McGregor Poll; Capt Fred Olmsted; Capt Jim Meehan; Maj Ronald W. Gibbs; Lt-Col Greg Swanson; Lt-Col Lanny Lancaster, who provided an enormous amount of information on tactics; Maj Harry Edwards; Brig-Gen Robin Olds (USAF, Retd); Lt-Gen John J. Burns (USAF, Retd), whose information and insight were invaluable; and Sgt Rondell M. Beach.

Navy and Marine Corps officers who were equally helpful include Capt Gene Tucker, who gave a broad view of combat from the carriers; Cdr Randy Cunningham; Lt-Cdr William Driscoll, who provided some excellent tapes; Lt-Col L. G. Karsch; Cdr James Spencer; Rear-Admiral Harry Gerhard (USN, Retd); and Maj-Gen Hal Vincent (USMC, Retd).

At McDonnell Douglas "Beaver" Blake introduced me to a host of people whose experience ranged from work on the preliminary Phantom projects to service as top company executives. The president of the McDonnell Aircraft Company, Donald Malvern, was generous with his time and was able to put the Phantom programme into perspective for me. The amazing Herman D. Barkey, who kept the Phantom being the Phantom during the long period when many forces were seeking changes, was invaluable. Others who gave freely of their time and insight include Harold D. Altis; Dan Freeburg; D. D. Clark; H. Perlmutter; Thomas Plein; W. J. Blatz; A. Boyd; G. Gregurec; J. Kohoutek; Dave Freidman; Jack Krings; F. Bloomcamp; R. A. Schoppman; Carl F. Tarricone; Ray Juergens; Gordy Graham; Walter House, a fellow historian; John "Jack" Harty, who provided some marvellous material; Lon Nordeen, a powerful writer in his own right; the wonderful Harry Gann, who has done so much for aviation photography and who gets more flying time than almost anyone; and members of the public relations staff, including Craig Smith, Tim Beecher, Jeff Fisteris, Gordon Le Bert (who provided many files), Doree Martin, Rose Dyer and Sharon Farrell.

I received a great deal of help from Brig-Gen Abel's Air Force staff, including Lt-Col Eric M. Solander and Capt Sam Brown. Also helpful were Susan Simpson of DARVA; Jack M. Woods of the Ogden ALC; Fred Johnsen, information officer at McChord AFB and an able historian; and Bob Laur. Air Force historian Dr Richard Kohn was as supportive as ever, as was Dr Cargill Hall of the Albert E. Simpson Research Centre, Maxwell AFB, Alabama, where Gerard Hasselwande did some excellent work for me.

In a similar way, many people in the Navy and Marine Corps were most helpful. Col Maas put out the word that I needed information, as did Capt Mac Snowden; others included Roy Grossneck, Hal Andrews, J. F. Skelly and one of the early architects of the Phantom programme, George Spangenburg.

In the media, William Gregory and Robin D. Schauseil of *Aviation Week and Space Technology* were as helpful as ever, as were William Schlitz of *Air Force Magazine*; Robert Williams of *American Aviation Historical Society Journal*; Gordon Swanborough of *Air Enthusiast*; Charles Cain of *Air Pictorial*; and Wolf I. Blitzer, Hirsch Goodman and Javad Khabkhaz. Bernard Thouanel of *Aviation International* provided some much needed photos.

Amongst the historian, buff and photographic fraternity, I owe a great deal to old friends Jay Miller, Vic Seely, Fred Dickey, Bob Lawson and Dave Menard, all of whom give generously to fellow writers all the time; and to new friends Nick Williams, Harold Stockman, Paul Swendrowski,

Norm Taylor, V.J. Van den Berg, David W. Schill, Mike Skinner, Richard Copsey, Edward Fenner, Don Linn, Eric Renth and Jeff Norgroove.

Keith Ferris was, as always, generous with his talent.

Once again, I am indebted to everyone who helped, and only wish I could convey to each of you how much your efforts meant. I want also to thank my wife Jeanne and daughter Peggy (the other kids are out of the nest) for putting up with the months of toil and the endless preoccupation with the book. Thank you all, everyone.

Walter J. Boyne
Alexandria, Virginia

# Introduction

The McDonnell Douglas F-4 Phantom is a remarkable aircraft, built, flown and maintained by remarkable people. I have been very fortunate in having had their total co-operation.

The story of the F-4 in combat is a long and complex one. I have chosen to write it in a manner which presents the views of the men who flew it, and of the commanders who used it as an extremely valuable tool. Inevitably, the Vietnam War dominates the narrative, and the American airmen's recollections generally express frustration and disappointment at the restricted manner in which the Phantom, and air power in general, were employed in the conflict in Southeast Asia. I know full well that there were many factors which prompted these restrictive rules and procedures, including the desire to keep China out of the war and the need to respond to a small but vocal part of the American public which wanted the United States out of Vietnam at any cost.

If the war had not been made so costly and protracted by these considerations, and if in the final weeks of 1972 the rules had not been stood on their head in an effort to stem the onrushing tide of North Vietnamese troops, it might have been possible to write a more balanced account. Indeed, I can understand well how policies evolve and change over time. But in describing the combat career of the Phantom, in charting the dogfights, the raids, the exchange ratios, I felt it necessary to remind the reader regularly of the artificial constraints placed upon the aircraft's use. In any event, the opinions are my own, based on many hours of discussion with many of the principals of the time. I hope that I have faithfully represented them, as faithfully as they flew their aircraft in a war that no one wanted.

The F-4, in all its many versions, is a historic machine. The world will probably never again see an aircraft which maintained such a decisive edge for so long. But, eventful as the Phantom's career has been, the story depends ultimately on its human characters. This, more than anything else, is what I wish to convey.

# 1. A taste of combat

The McDonnell Douglas F-4 Phantom II is the classic modern fighter of the Free World. A total of 5,195 have been built, including 138 in Japan, for use by 11 countries. Many of these aircraft have been rebuilt and fitted with new equipment, for the basic performance of the original Phantom II* was so advanced for its time that it still compares favourably with that of aircraft currently in production. When updated with computerised weapon systems and electronic countermeasures, it is competitive with the best the Eastern Bloc can offer. It is still the principal tactical photo-reconnaissance and ground-attack aircraft of many air forces. In its original role of air-superiority fighter it has been superseded but not entirely supplanted by later aircraft, and under the right conditions it can be the equal of aircraft costing perhaps five times its original price.

Almost unwittingly, by a happy combination of brains, luck, determination and good timing, the McDonnell team delivered in 1958 an aircraft which leaped a generation in development. To the uninitiated, the XF4H (almost a misnomer, for the very first Phantoms were true production aircraft, not experimental prototypes) looked "all wrong", with its drooped snoot, cranked wings, sagging anhedral tail surfaces and semi-submerged missiles. The cavernous, slab-sided holes flanking the cockpit looked more like airbrakes than intakes.

*Phantom II is still the correct designation of the F-4, but because there is no danger of confusing it with the original FH-1 Phantom it will be referred to simply as the Phantom from this point on.

**Above:** This McDonnell Douglas publicity shot was issued in 1981, when the Phantom's air-combat kill total stood at 278, including 108 MiG-21s. *(MDC)*

**Left:** F-4C launches a Sparrow. This radar-homing missile survived a bad start in Vietnam to become a proven MiG-killer in the hands of USAF and US Navy Phantom pilots. *(MDC)*

**Right:** Page from a US Navy tactical manual showing how the Phantom's superior performance in the vertical plane could be used to defeat a tighter-turning opponent. *(US Navy)*

# ROLL-AWAY MANEUVER

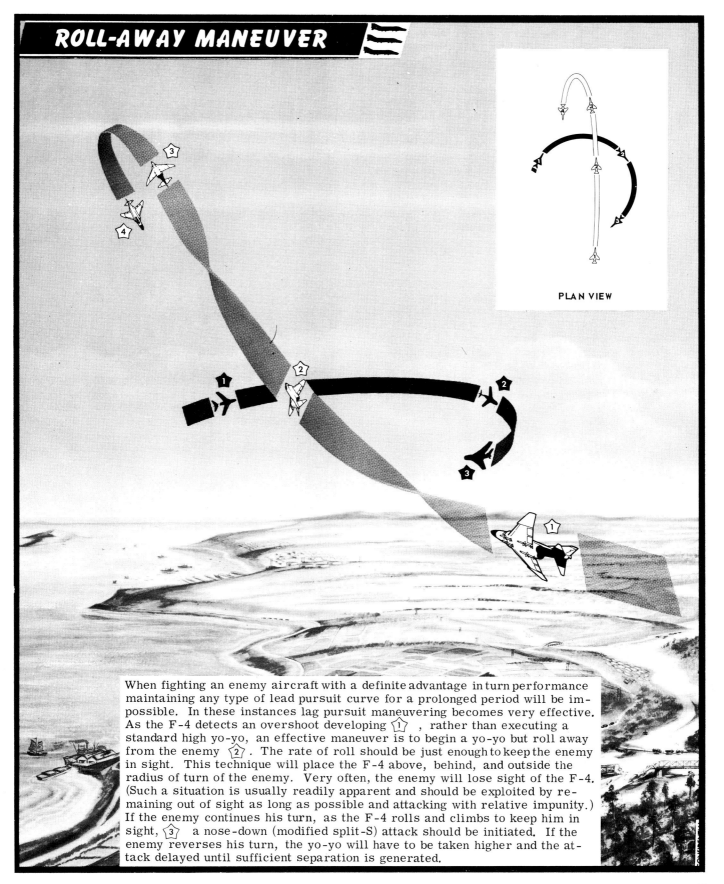

PLAN VIEW

When fighting an enemy aircraft with a definite advantage in turn performance maintaining any type of lead pursuit curve for a prolonged period will be impossible. In these instances lag pursuit maneuvering becomes very effective. As the F-4 detects an overshoot developing ①, rather than executing a standard high yo-yo, an effective maneuver is to begin a yo-yo but roll away from the enemy ②. The rate of roll should be just enough to keep the enemy in sight. This technique will place the F-4 above, behind, and outside the radius of turn of the enemy. Very often, the enemy will lose sight of the F-4. (Such a situation is usually readily apparent and should be exploited by remaining out of sight as long as possible and attacking with relative impunity.) If the enemy continues his turn, as the F-4 rolls and climbs to keep him in sight, ③ a nose-down (modified split-S) attack should be initiated. If the enemy reverses his turn, the yo-yo will have to be taken higher and the attack delayed until sufficient separation is generated.

Amazingly, the Phantom took air forces further than they were prepared to go: it had a speed and altitude capability in excess of that required by contemporary combat concepts, one that would only be matched by the next generation of fighters. The limitations of men and missiles drove air combat down to relatively low altitudes – from the deck to perhaps 25,000ft being the rule – and most fighting was done in the 500-900mph bracket. It did not at first dawn on the McDonnell engineers, nor the services, nor fortunately the enemy that the performance potential of the Phantom could be translated into a new style of combat: "energy manoeuvrability," in which the classic horizontal turning dogfight would be abandoned for manoeuvres in the vertical plane. The thrust-to-weight ratio of the F-4 could be used to boost it up and down in yo-yo and barrel-roll manoeuvres which would effectively counter the tighter horizontal turning capability of its lightweight MiG opponents. This attribute more than made up for the F-4's greater size and gross weight, which were required to provide the range, loiter time, weight of armament and variety of equipment necessary for its multi-mission role.

Long before these concepts were proved in action, however, the F-4 had sold itself to the world by establishing a series of records that at first seemed scarcely credible. In a breathtaking 26 months, newly minted F-4s set the following marks:

| Category | New record | Date | Aircraft commander |
|---|---|---|---|
| Absolute altitude | 98,557ft | Dec 6, 1959 | Cdr L. E. Flint Jr |
| 500km closed course | 1216.8mph | Sep 5, 1960 | Lt-Col T. H. Miller |
| 100km closed course | 1390.24mph | Sep 25, 1960 | Cdr J. F. Davis |
| 3km closed course | 902.8mph | Aug 28, 1961 | Lt H. Hardisty |
| Absolute speed | 1606.3mph | Nov 22, 1961 | Lt-Col R. B. Robinson |
| Time to height | 9842.5ft (3,000m) in 34.52sec to 98,425ft (30,000m) in 371.43sec | Feb 21, 1962 to Apr 4, 1962 | Various |

These records served notice to the world that the Phantom had made all other fighters obsolete, and forced leaders of air forces everywhere to start thinking about how to use (or oppose) an aircraft of such awesome potential. The old rules no longer applied, for here was a multi-purpose aircraft that promised to outdo contemporary single-role aircraft in each of their specialities. It proved to be a better interceptor than the Convair F-106, better at ground attack than the Republic F-105 (although loyal "Thud" drivers will contest this), a better fleet defence aircraft than the Vought F-8 Crusader and its derivatives, and, it can be argued, a better reconnaissance type than its elder brother the F-101.

This versatility led to a curious situation in which the Phantom's potential far exceeded the ability of various air forces to train crews to exploit it. In time these air arms even came to believe that multiple-mission aircraft were less desirable than specialised aircraft with crews dedicated to one role or another. As we shall see, various services approached the challenge differently, with different results. The USAF tried to funnel all of its pilots, whatever their specialisation, through courses which nominally made them fighter pilots but which in many instances left them inadequately trained and motivated to use the F-4 to the full. The US Navy, by contrast, trained each crew to fight as a unit in a selected role, and as a result obtained better use of the Phantom. The Israeli Air Force (*La Tsvah Hagana Le Israel/Heyl Ha' Avir*) seized upon the F-4 as a means of multiplying its air power, and at the same time kept an élite

**Left: F-4C of the 557th Tactical Fighter Squadron, 12th Tactical Fighter Wing, based at Cam Ranh Bay during the Vietnam War, drops a napalm bomb on a shoreline target.**

**Above:** The USAF soon saw a need for guns on the Phantom. Pods with 20mm Gatling-type cannon were installed but were never as satisfactory as an internally mounted gun. *(MDC)*

**Below:** Although the Navy was the first to buy the Phantom, the Air Force ultimately bought more (2,582 against 1,264) and experimented more widely. This is a Navy F-4B on loan to the USAF. *(MDC)*

of aircrews constantly annealed in the fierce fires of combat or equally rigorous training. Where ageing French Vautours had been pushed beyond all sensible limits, the F-4s could range in relative comfort across the territory of Israel's encircling enemies, performing such feats as the laying of sonic booms on Cairo, the bombing of the Iraqi nuclear reactor, and the remarkable obliteration of Syrian missile defences in the Beka'a during the 1982 Lebanon conflict.

What made the Phantom more than could be absorbed by most of the aircrews of the twenty air forces which have used it? It was a combination of advanced engineering, superlative engines, extraordinary structure, remarkable power-to-weight ratio, adequate wing area, computers and integrated avionics that were ahead of their time, excellent armament, and a military/industrial team that brought forth fixes, modifications and updates as required to stay ahead of the game. If the user service could add superior aircrew training and high motivation, McDonnell's Phantom became a world-beater.

The Phantom also came to dispel one of the great myths of the 20th century, that of Eastern Bloc air power. Despite the fact that the MiG family of fighters and its Sukhoi brethren are excellent aircraft, despite the fact that combat has usually occurred over communist territory, despite the fact that MiGs are built in numbers undreamed of by Western manufacturers, despite the fact that they were deployed in a ground control environment superior to anything in the Free World, and despite the fact that they were supplemented by the most formidable anti-aircraft artillery and surface-to-air missile systems in history, Soviet philosophy reduced them to *Potemkin* weapons, great for show but employed in a fatally ineffective manner.

Whatever the origin of the MiG pilots – Soviet, Chinese, North Vietnamese, North Korean or Arab – they have usually been unmotivated and relatively unaggressive. Some of this stems from the fact that communist air forces, and those other air arms to which they have supplied aircraft and training, have been trounced every time they have been opposed by Western-trained forces.

The Phantoms, as we shall see, were often deployed in the most unfavourable circumstances, far from supply sources, in the worst of weathers, and constrained by unrealistic procedures and rules of engagement. Their MiG opponents, by contrast, sortied only when the weather suited them and engaged only when combat conditions were deemed perfect. But they went out in the full expectation of being beaten, and most often they were.

This argument is of course apparently refuted by the decline in kill/loss ratio experienced by US forces in Vietnam by comparison to those of Korea and the Second World War. Later analysis will show that the ratio was in fact extraordinarily good in view of the politically inspired rules of engagement imposed upon the American air forces. The rules gradually changed and were ultimately reduced to the point where air power could take effect. But before then

it was as if Dowding had been forbidden to use radar during the Battle of Britain, or if Rommel had been required to drive his tanks only on certain roads at preannounced times.

In no other combat were surprise and uncertainty so willingly forfeited to the enemy. Whenever before had most of the enemy's territory been given such sweeping sanctuary that attacks could only be carried out along narrowly defined routes, with defences clustered along each side? Whenever before had all the key targets been excluded, decisive weapons forsworn, and "bombing holidays" given? Instead of applying force at the source of the enemy's strength, the Americans dissipated their efforts across a friendly countryside, dimpling South Vietnam and Laos with countless craters as multi-million-dollar fighter-bombers attacked elusive, hardy, brave individuals clad in black pyjamas and carrying 30kg sacks of rice.

The cumbersome restrictions were however offset by an outpouring of resources which permitted some chance of success and the best possible hope of survival. President Johnson, and later President Nixon, may have hampered the execution of the air war but they provided the necessary aircraft, equipment, munitions, ships and personnel in ample quantity. Backing up the offensive effort were sophisticated radar and communications systems and the most elaborate rescue facilities ever created. Similarly, while the rules of engagement may have obliged strike forces to fly certain predictable routes and to forgo attacks against obvious targets, tremendous efforts were made to develop the electronic countermeasures, "smart" (guided) bombs and other equipment needed for operations under hitherto impossible conditions.

But in the final analysis, when it came to aircraft-versus-aircraft, pilot-against-pilot combat, the North Vietnamese Air Force was deficient. Russian-trained, the North Vietnamese pilots lacked aggressiveness and teamwork, with no notion of mutual attack and defensive techniques. There were individual exceptions, but for the most part the North Vietnamese fliers seemed to rely on hit-and-run tactics despite the fact that they possessed superior dogfight aircraft and were operating within one of the world's most well developed and best integrated combinations of radar, anti-aircraft artillery, surface-to-air missiles and fighters.

Before moving on to a preliminary sample of Phantom combat reports, it is well to recall that many different missions have been carried out by the type. All too often the Phantom's wars are described only in terms of F-4/MiG encounters in which men, machines and missiles are thrown together for a few moments of gut-wrenching anxiety which end quickly in victory, defeat or stalemate. However, there were perhaps fifty other types of F-4 combat mission flown in the Vietnam War: some apparently less glorious, some obviously more dangerous, but all definitely going in harm's way. These included MiGCAP, ResCAP, close air support, Fast FAC (forward air control), landing-zone extractions, Spectre escort, Night Owl FAC, sensor delivery, mining, flak suppression, chaff dropping, B-52 escort, air defence alert, laser-guided bomb delivery, and electro-optically guided bomb delivery. To this list must be added the various missions developed by the Israeli Air Force.

Just as the tenor of combat varies, so does its reporting. Contemporary official combat reports from USAF or USN

**Above: Sometimes overlooked in the heat of combat was the work of the RF-4Cs. Their contribution to the air and land battle was invaluable, and the Phantom became one of the greatest reconnaissance types ever. This is a McDonnell test aircraft pictured before its first flight in RF-4C configuration, having been converted from F-4C standard.** *(MDC)*

**Left: The shadow of an RF-4C crosses a Vietnamese elephant grazing area. Note startled water birds rising.** *(USAF)*

**Below left: Like all pilots in all wars, the young Americans in Vietnam liked to give their aircraft distinctive names. This F-4C, 64-851 of the 557th TFS (the "Sharkbaits"), was stationed at Cam Ranh Bay during some of the most difficult times of the war.** *(Walter D. House)*

crews are usually hard-edged, professional commentaries on the techniques used, couched in intricate jargon and laced lightly with the humour of victory. Later assessments by the same people are more reflective, not so fast-paced and better balanced overall.

Reports from Israeli sources tend to be colourful but heavily censored. It is difficult to become emotionally involved with "Captain A." or "Major R." of "a certain squadron," but for good reasons Israeli security rules are rigid. Still, the Israeli Air Force has had the good fortune to be able to use the Phantom to the maximum, following realistic rules of engagement, and has established the highest kill/loss ratios in modern warfare. Some Israeli sources believe in fact that the importance placed on high exchange ratios has actually worked to the detriment of their air force. They feel that with air superiority already clearly established – recent exchange ratios have reached 40 or 60 : 1 – the IAF should devote more resources to developing anti-missile and flak-suppression tactics to avoid the terrible losses of the War of Attrition and the Yom Kippur War. Events in the Lebanon in 1982 suggest that this school of thought may have begun to prevail.

"You love a lot of things if you live around them, but there isn't any woman and there isn't any horse, not any before, nor any after, that is as lovely as a great airplane. And men who love them are faithful to them even though they leave them for others. A man has only one virginity to lose to fighters, and if it is a lovely plane he loses it to, there his heart will ever be." These words by Ernest Hemingway describe perfectly what motivated Capt Fred Olmsted USAF during his 20 months of combat experience flying Phantoms stationed at Udorn Royal Thai Air Force Base, Thailand. Olmsted now flies Boeing 727s for Eastern Air Lines, but during 1971 and 1972 he was among a small cadre of seasoned F-4 pilots who managed to observe the adverse rules of engagement and still came out on top. On March 30, 1972, he was on his 300th combat mission, flying with the Panther Pack (13th Tactical Fighter Squadron, 432 Tactical Reconnaissance Wing), with Capt Gerald R. Volloy as his weapon systems officer (WSO). Here is Olmsted's brief account of his first MiG kill:

"My first MiG-21 kill occurred on March 30, 1972. Our covert intelligence sources had uncovered a plan by the North Vietnamese Air Force whereby three MiG-21s would trap a 'slow mover' AC-130 gunship north of the DMZ [Demilitarised Zone]. Consequently I was put on full scramble alert along with two other highly experienced F-4 pilots. Within an hour of the intelligence guesstimated time of scramble, we took off with orders for a max afterburner climb and a full military power dash to the North Vietnamese/Laotian border.

"The US Navy took us under its radar control and vectored us to the vicinity of the three blue bandits – the MiG-21s. I engaged one of the MiGs in a low-level, night-time, high-speed series of turns. Whenever my back-seater would lock on, the bandit would turn towards the water; the Navy would then lock on with its shipboard missiles, and the MiG would turn back to my onrushing Phantom. [The NVAF's sophisticated electronic detection equipment, both ground-based and airborne, often dictated its choice of tactics and consequently the nature of the air battle.] The MiG finally turned head-on to me at a closure speed of

Left: F-4Ds of the 12th TFW line up at the fuel pits following missions over Vietnam. Photograph taken at Phu Cat, Vietnam, on May 25, 1971. *(Norman E. Taylor)*

Centre left: Dissatisfaction with pod-mounted cannon led to the design of an internal gun. This is perennial test ship 12220 with mock-up gun installation. *(MDC)*

Bottom left: F-4Es with gun installed lent themselves readily to the sharktooth treatment. The M-61A1 20mm gun could be fired at either 6,000 or 4,000 rounds/min, selectable from the cockpit. Muzzle velocity was 3,380ft/sec, and 44lb of projectiles could be thrown in a two-second burst. *(Jay Miller)*

greater than 1,000kt. I fired a Sparrow radar-guided missile. The first missile didn't guide well, so I fired again. Both of us in the F-4 saw the missile guide and in a second or two saw a tremendous fireball as the Sparrow impacted the MiG head-on. The Navy then confirmed that the bandit had suddenly disappeared from its radar scopes."

Capt Olmsted's second victory came on April 16, 1972, and in many ways is a textbook example of how to fight air-to-air in the F-4. Olmsted served as wing tactics officer (one of the more desirable "additional duty" jobs in a fighter unit) and incessantly extolled the vertical fighting performance of the Phantom against the quick-turning MiG-21s. He recalls: "I was leading Basco flight, MiGcapping west of Hanoi, with the task of keeping MiGs away from our F-4s inbound on a bombing run. Soon after entering our assigned MiG orbit, my back-seater, Capt Stu Maas, identified two blue bandits closing in on us from 12 o'clock and a range of 20 miles. We tracked them down the radar scope and had perfect head-on shots at both of them; I hesitated, however, fearing that it might be friendlies straying into our area. Alas, two silver MiGs flew immediately past us, between my element of two and Maj Dan Cherry's element. I reversed the four-ship tactical formation, rolled out of a 180° turn, and visually spotted the two bandits dead ahead at a range of two to three miles. I lit the afterburners, pulled the nose up and watched the lead MiG roll inverted and split-S in front of my nose, leaving his less experienced wingman to do combat alone. The MiG broke hard left (4g turn approximately), inviting me to pursue horizontally. I answered, however, by initiating a high-speed barrel-roll attack up and outside of the MiG's turn, followed by a high-speed yo-yo to the inside. Each time the MiG broke hard to avoid the Phantom, I countered vertically (barrel roll and/or high-speed yo-yo); as the fight progressed through five or six high-g manoeuvres, I had closed to minimum missile range and I loosed an AIM-7E Sparrow. The missile hit the MiG's high wing (it was in a left turning spiral) and sawed it in half, but the MiG pilot didn't eject nor did the plane catch fire. I closed accordingly and fired another Sparrow, which guided perfectly. The Sparrow impacted the MiG in the canopy and the bandit exploded into two huge fireballs."

We'll hear more from Fred Olmsted later.

The Israeli Air Force has problems very different from that faced by the Americans in Vietnam. Instead of having to concentrate its efforts on narrow strips of highly defended territory, the IAF must defend against assaults from every point of the compass and be prepared to launch attacks in response. Israel is at most roughly 260 miles long and about ten miles wide at its narrowest. A MiG-25 can penetrate, overfly and depart in minutes. It is surpassingly difficult even to run training exercises and be certain that the airspace of neighbouring – and uniformly hostile – countries is not violated. As will be shown later, Israel overcomes these handicaps by intensive training of a carefully selected force. Its combat crews are led by veterans of as many as three wars, and its political and military leaderships are generally in close accord when it comes to the conduct of war.

Israel abides by the rules of engagement used by "civilised" combatants the world over – careful avoidance of civilian casualties, no attack unless the target is positively identified, and so on – although the Lebanon conflict in 1982 put severe strains on any Israeli pilot who tried to stick to the letter of the law. The IAF is not hampered by concerns for its opponents' philosophy or state of mind, nor for world opinion: if an attack is necessary it is carried out with all the surprise that can be achieved and all the ferocity that can be mustered. Israel's wars are fought very close to home, and an enemy repelled on one front may be immediately succoured from another quarter. And there always remains the possibility that the country will simply be overrun.

The Israeli fighter pilot belongs to the élite of the élite, and despite the acquisition of the formidable General Dynamics F-16 and the McDonnell Douglas F-15 the Phantom pilot is especially highly regarded. The well informed Israeli public knows how demanding it is to operate the Phantom in its multiplicity of roles, especially against the Arab complexes of surface-to-air missiles. The IAF will not reveal the number of Phantoms lost out of the 216 supplied by the United States, but the type still constitutes the Air Force's leading edge. From September 7, 1969, when the first F-4Es brought a new dimension to Israeli air power, to the summer of 1982, when F-4s armed with native-built electronic gear and weaponry helped destroy the labyrinth of Syrian surface-to-air missile systems in the Lebanon, the Phantom has been both shield and sword.

Oddly, little of the savagery of the fighting comes through in *Biaton Heyl Ha'Avir*, the Israeli Air Force journal, which provides most of the insight available to outsiders. The

Above: F-4E 67-315 of the 469th TFS, 388th TFW, en route to target area over Vietnam. Note ordnance and vapour condensation in the core of the tip vortex. *(Don Logan)*

*Above:* An F-4D of the 389th TFS, 12th TFW, based at Phu Cat, prepares to depart on a mission to North Vietnam. It is armed with six 500lb high-drag bombs, three napalm tanks attached to the centreline station, and two Sparrows. Photograph taken on April 17, 1971. *(Norman E. Taylor)*

*Below:* The sweetest moment! Capt Frederick S. Olmsted Jr climbs down from the cockpit while a ground crewman gets ready to paint on the star for his April 16, 1972, victory. Olmsted, flying with the 13th TFS, 432nd TRW, had just downed a MiG-21 with a Sparrow. *(Fred Olmsted)*

following account by Yakir Alkriv, "Cleaning the Skies over Sharom," is typical.

"The first air battle of the Yom Kippur War has already entered history as a rarity. This is not only because it was one of the few combats that occurred over Ophir base, and not only because seven enemy aircraft were shot down by two of ours. It is special because of the events of the battle itself.

"Lt-Col S. was one of the two pilots who participated, and he relates: The war was a complete surprise to me. On Friday I was at my base when someone who was in readiness at Sharm asked me to replace him. I was relatively new on the Phantom and I agreed to replace him; the same day I went to Sharm.

"The readiness was absolutely ordinary.... I remember that we had a good time in Eilat until late in the evening; the last thing we thought about was war. We woke up Saturday at ten in the morning and found that Col (Reserves) Jack Nevo* was arriving to take command of the base.... The situation was quiet, so why so sudden?... Jack landed at one in the afternoon. The second sentence that he spoke I remember to this day: 'The Egyptians are ready, apparently, to begin war.' Before he had finished speaking, the warning sirens started and the war began for me.

"We were a pair in readiness, N. in Number 1 and I as Number 2. We ran to the aircraft, started up and waited for the controller's orders, which were delayed by something. We shouted at him to authorise our take-off. The tension was great. The controller said yes, and we scrambled into the air like crazy. We turned immediately south-west. When I dropped a wing to look around, I saw holes from bombs close to the runway we had just taken off from a moment ago.

"My navigator, L., and I immediately began to search for enemy aircraft and we found a flight of MiG-17s heading towards the base, fully armed. The pursuit began immediately. We thought to ourselves that only yesterday we were having a good time in Eilat and today we were pursuing a flight of MiGs who are surely trying to erase us from the map.

"I immediately took the lead and ran ahead. The MiG pilots looked around and began to flee towards Egypt. L. and I got behind them quickly and immediately launched our first missile. The missile disappeared somewhere into the sky, apparently suffering from a technical fault. But the second missile scored a direct hit and the second MiG exploded into pieces in the air. His wingman tried to dive low to flee but an additional missile caught him and he crashed into the sea. Now our spirits were high. L. and I searched for additional prey and it wasn't long in coming. Two MiG-21s came at us in a large circle to release bombs on the base. Before they found out what was happening we got behind them in a firing position. This caused a problem for both them and us. Our problem was that they were exactly over the base: if we shot them down on the runway there would be no other place to land. So we waited quietly until they had crossed the runway and then we launched a missile at one of them. Air-to-air missiles always perform some manoeuvre

*Col Nevo shot down one and possibly two MiG-15s while flying a Mystère IVA during the 1956 campaign. He later became the first commander of the IDF/AF's Super Mystère squadron.

Right: Painting up the 50,000th bomb to be dropped by the 12th TFW. Customs like this seem to endure, whatever the time or the war. *(USAF)*

Below: The Navy did not develop a Phantom with an internal gun, preferring to stay with the Sparrow (shown here) and Sidewinder. *(MDC)*

Bottom: Battle damage was frequent in the Vietnam War, and the US Navy's on-board repair facilities were limited. This Phantom was sent to the Naval Air Rework Facility, North Island, to have extensive fire damage made good. *(US Navy)*

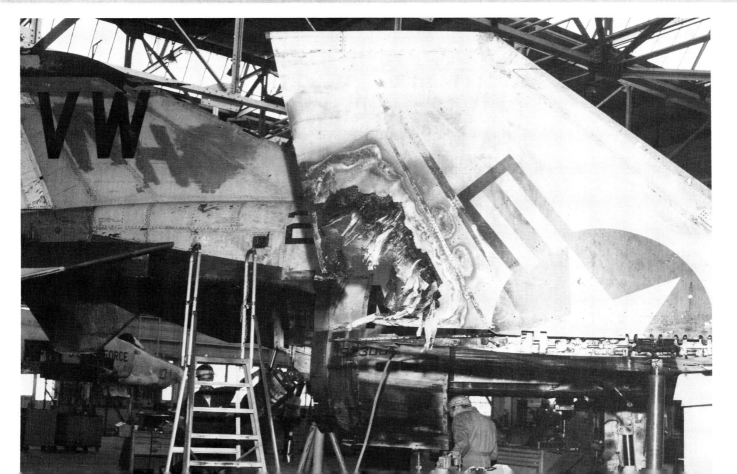

on their way to the target. But this missile performed a corkscrew so large that we were sure that it was going to hit the ground-controlled interception position on the base, located a little to the side of the course of the MiGs. Suddenly the missile stopped its crazy evolutions and – Boom! – exploded the MiG into small pieces.

"At this point there occurred the strangest thing in my flying career. In trying to escape, the second MiG flew low, really hugging the ground. I pursued him into the wadis that surrounded the base but I couldn't manage to put the sight on him to launch a missile. Then he crossed the shoreline and flew parallel to it. He descended lower and suddenly a wall of water rose behind me. I was sure he had dived into the sea but to my stupefaction I saw him manoeuvre and continue to fly. At 500kt he touched the sea and exited from it, and as if this wasn't enough he performed the manoeuvre three times!

"But the story isn't finished. On the way home a flight of MiG-21s came at us. We still had two missiles and many cannon shells but not enough fuel for air combat. And so despite a great temptation – I remember the disappointment to this day – we abandoned the MiGs and returned to land on the damaged runway. When we landed we found that more pilots had arrived with heavy armament loads. After a lot of trading we divided the missiles and shells between us and began continuous missions for the rest of the day and night.

"The height of the story came immediately after that first landing. Base commander Jack Nevo came up to us and asked 'How many?' I said: 'Seven. We had three and Number 1 had four.' Jack simply didn't believe me. If a combat veteran like Jack doesn't believe you it's a sign that you've downed a lot of aircraft."

Later chapters will include more combat stories, covering the full spectrum from reconnaissance to bombing. But first, how did the Phantom come to be so providentially capable?

Left: Proof that it actually happened. It was May 10, 1966, when Lt Greg Schwalbert of VF-14 launched with his wings still folded. The Phantom weighed 34,000lb on take-off, outside air temperature was 85°, and wind over the deck 33kt. The external load was jettisoned and the aircraft landed safely after a 170-180kt approach to NAAS Leeward Point, 59nm from the carrier. While this is not a combat shot, it serves to show just how powerful the Phantom is. *(MDC)*

Below left: Maintenance below deck was never easy, hence the general look of aggravation at something wrong with the refuelling boom.

Above: A shamrock-bedecked Phantom from VFMA-333 ready to launch.

Below: An Israeli F-4E taxies past a Fouga Magister somewhere in Israel. As usual, the censors have blotted out squadron insignia. *(MDC)*

Left: moving into position on the catapult. The name written on the wheel-well door – "Spaced", with its connotations of drug use – must have escaped official scrutiny.

Above: Somehow symbolic of Israel's beleaguered position in the Middle East, an F-4E streaks over a barbed wire entanglement.

Right: The enemy below: a North Vietnamese MiG-21 Fishbed photographed near Hanoi. *(USAF)*

# 2. Ghostly lineage

Reading company reports is not ordinarily a thrilling way to pass the time, but a glance at the annual returns of the McDonnell Aircraft Corporation, as it was originally known, is highly revealing of James Smith McDonnell, "Mr Mac," and his methods. The firm was incorporated in Maryland on July 6, 1939, and started business in St Louis with two employees, McDonnell and his secretary, Lou Ritter. McDonnell had amassed $30,000 in savings, to which was added a further $165,000 from friends and relatives. By October 15, 1939, he had acquired 15 engineering personnel and was ready to do business – any sort of business – relating to the design, construction or repair of aircraft and parts.

From the beginning McDonnell was both prudent and extravagant with funds: he didn't waste a penny, but would spend whatever was needed to improve his facilities and equipment. Similarly, he would demand tremendous efforts from his employees and then reward them handsomely. He was that rare combination in aviation, the engineer-businessman. He was a brilliant, far-seeing engineer whose quest for the outer limits of performance never interfered with his understanding of the profit-and-loss statement or the balance sheet.

But above all the key to McDonnell's success was his ability to capitalise on events, turning everything to the advantage of his firm from the beginning of the Second World War to its end, from the introduction of jet engines to the beginnings of the space age. There were misadventures, to be sure: his early association with Constantine Zakhartchenko on the Doodlebug project for the Guggenheim Safe Aircraft Competition carried over into the war years and after as an infatuation with helicopters and convertiplanes. There were technical successes but they won no production contracts, and nor did a well engineered foray into what has become the executive jet business. The main effort of the firm went into successful ventures, however, and failures such as the Little Henry, Whirlaway and 220 transport projects received nothing more than initial development funds. Just as he had an unerring instinct for a winner, McDonnell also knew exactly when to cut his losses.

McDonnell Aircraft went from strength to strength on the basis of its fighter aircraft programme. The 1940 annual report shows that the firm offered no fewer than eleven designs to the Army and four to the Navy in its first year of existence. Though the only tangible result was a single $3,000 award from one Army design competition, McDonnell had secured over $3 million in contracts for parts and engineering, and the company was on its way.

Mr Mac's persistent efforts to woo the US Army won him a contract in September 1941 for the lovely-looking XP-67, the second aircraft to be built under the McDonnell name. The company's energetic performance on the XP-67 was assisted by a $15 million contract to build Fairchild AT-21 bomber trainers in a new plant at Memphis, Tennessee, and this successful effort gave the fledgeling firm experience in mass production.

The brilliance of the XP-67 design was evident at a glance: it featured wing/fuselage blending of the kind applied years later in more sophisticated form to the General Dynamics F-16. Striking in appearance, the XP-67 was however handicapped by its never adequately developed Continental XI-1430 inverted-vee liquid-cooled engines, and by an insufficiency of wind-tunnel work. Despite its advanced if not thoroughly refined aerodynamics, heavy firepower of six 37mm cannon, pressure cabin and use of exhaust to augment thrust, it was not acceptable to the US Army Air Forces. An in-flight fire on September 6, 1944, put a stop to tests, and the project was abandoned.

But McDonnell's now obvious ability to innovate had not gone unnoticed. Wartime pressures to maintain production of existing proven designs and engines kept Pratt & Whitney out of the beginning of jet engine development, and permitted contracts for experimental jet fighters to go to firms like Bell Aircraft and McDonnell. In January 1943 McDonnell had received a contract that led to the design of the XFD-1 Phantom carrier-based fighter (not to be confused with the Phantom II, subject of this book). Kendall

**Above: James Smith McDonnell started his career with Huff Daland, which for a long time produced America's ugliest aircraft. Then the company came upon the Fokker formula and started building D.VII lookalikes like this AT-1. When McDonnell started work in 1924 there were only 200 aeronautical engineers in the US industry. When at last he retired he was pre-eminent amongst America's hundreds of thousands of aerospace specialists.**

**Top:** The ancestry of the Ford Tri-motor has never been absolutely established, though at least it is now generally accepted that Bill Stout did *not* design it. The "Tin Goose" was in fact the work of a team which included the young James McDonnell.
*(US Air Force Museum)*

**Above:** The first aircraft to bear McDonnell's name was the Doodlebug, his attempt to win the Guggenheim Safe Aircraft competition. He worked with Constantine Zakhartchenko and others on the aircraft, which offered remarkable performance but was not able to overcome a series of normal test programme difficulties in time to win. Zakhartchenko would later be a proponent of helicopters and convertiplanes at McDonnell.

**Right:** McDonnell became chief engineer for landplanes at Glenn L. Martin and was the architect of two hot-rod bombers, the Maryland and Baltimore. Here famous test pilot Sam Shannon poses beside the cockpit of the slender Maryland, which was designated A-22 by the Air Force. *(NASM)*

Below: The Baltimore was an improved version of the Maryland, with larger engines and fighter-like performance. It was well liked by the crews (mainly French and British) that used it.
*(Glenn L. Martin)*

Below: Of the many McDonnell designs that have been called less than handsome, the XF-85 parasite fighter was certainly the most curious. Designed by the same Herman Barkey who would be largely responsible for the success of the initial Phantom II, the XF-85 was mercifully grounded before anyone was killed. The tiny jet fighter, with a wingspan of only 21ft 1.25in, was intended to protect the B-36 intercontinental bomber, being launched and recovered with the trapeze-like device shown here fitted to a B-29.

Right: The first aircraft from the newly formed McDonnell Aircraft Corporation was the XP-67, the result of a very sophisticated design which called for a blended fuselage and wing, and the maintenance of the aerofoil section through the fuselage and nacelles. Excessive drag and the troublesome Continental engines spelled the end of the programme, but both the Navy and the Air Force had been intrigued.

Above: McDonnell's willingness to work with the Navy in experimenting with the new jet engine during the war led to a contract for an aircraft that gave rise to one of the most successful series of fighters in history. The McDonnell XFD-1 Phantom (later FH-1 in production) was the first American jet to take off from and land on a carrier, flying from USS *Franklin D. Roosevelt* on July 21, 1946. Lt-Cdr James J. Davidson was the pilot.

Left: The Whirlaway (named after a popular racehorse of the time) hovers above an FH-1 Phantom. The Whirlaway was developed from the Platt-LePage line of helicopters under the tutelage of Constantine Zakhartchenko. *(NASM)*

Right: McDonnell XF-88 was a sophisticated penetration fighter with extremely clean lines, as this nose-on view shows. Though the XF-88 did not go into production, it prepared the way for the highly successful F-101. *(NASM)*

Below: The McDonnell F-101 Voodoo was the first supersonic fighter from the St Louis firm. It was beloved by pilots in all of its many versions, and 807 were built. First flown in 1954, the reconnaissance RF-101 version gave excellent service in Vietnam more than 15 years later. *(MDC)*

Above: Not a lovely aircraft and cursed at first with an inadequate engine, the McDonnell F3H Demon was ultimately well liked by its pilots. Along with the F-101, it prepared the way for the Phantom II. *(MDC)*

Right: Early Sparrows on the Demon. Missile technology did not match the hopes of the engineers for years. When the missile fighter concept was introduced, it was intended that they be employed against bombers, relatively easy targets. In practice missiles had to be used against agile, fast-turning fighters, and their performance had to be improved accordingly. *(MDC)*

Above: The very first and still unpainted Phantom II being manhandled in the shop by a group of men who would ultimately make a career building the aircraft. *(MDC)*

Perkins, formerly with Curtiss Wright, and Robert J. Baldwin were given the task of bringing the $4.4 million contract to fruition. This early work gave rise to principles of weight control, maintenance and production which today are still in use at McDonnell.

Drawing on XP-67 experience and production work on other projects, the Phantom proved to be moderately successful. Its achievement appears all the more remarkable when one considers the atmosphere in which it was produced. When McDonnell entered production with the FD-1 (later redesignated FH-1) on March 7, 1945, the war in Europe was obviously won and planners in the Army and Navy were already reducing and cancelling production contracts. Yet here was McDonnell, a young company which had never put an aircraft of its own design into production, about to take on giants like Douglas, Boeing, Lockheed, North American, Vought, Grumman and Curtiss. It seemed insane to compete with these giants, whose combined production rate was approaching 100,000 aircraft per year, whose assets dwarfed McDonnell's, whose engineering staffs were five to six times as large, and who had enough ready cash to buy and sell the newcomer. Moreover, many of these firms had custom and tradition on their side. Vought and Grumman had always supplied Navy fighters and had done a brilliant job in winning the naval air war. Their engineering staffs were considered to be the best in the world. But McDonnell would prevail, as was made plain on January 26, 1945, when the XFD-1 made its first flight.

Contrary to many previous reports, when Woodward Burke lifted off on that first flight in the first Phantom the aircraft was powered by two Westinghouse 19XB-2B engines rather than one. (Burke subsequently lost his life in the crash of the No 2 Phantom.) The XFD-1 resembled the earlier XP-67 in outline but weighed only one-third as much at a normal loaded 8,626lb. The Westinghouse engines (for a time called "Yankee" as a trade name) ultimately developed 1,600lb of thrust each, giving the Phantom a maximum speed of 487mph.

The tricycle undercarriage, dictated by problems associated with the jet efflux, was new for carrier work and some thought it might be unsuitable. But on July 21, 1946, all doubts were dispelled when the Phantom, piloted by Lt-Cdr James J. Davidson, made a series of take-offs and landings on the deck of the USS *Franklin D. Roosevelt* to become the first US jet-propelled aircraft to operate from a carrier.

The success of the Phantom came at a time when all other aircraft manufacturers were desperately trying to keep body and soul together, and spurred James S. McDonnell on to new efforts. Increasingly, Mr Mac assumed the role of the pioneering entrepreneur who could convince the Navy and, soon, the Air Force to buy new types before production of the previous one had finished. The production orders for 100 Phantoms had been quietly accompanied by a contract for the design, construction and flight test of the XF2D-1 (soon to be XF2H-1) Banshee, the beloved "Banjo" which would confirm McDonnell's growing reputation with the Navy.

By the end of the Second World War, McDonnell had manufactured 7,000,000lb of airframe and performed work to the value of more than $60 million. Even before the Phantom had flown from the *Roosevelt* McDonnell had obtained contracts for the radical XP-85 Goblin parasite fighter, the XP-88 Voodoo long-range fighter, and the Little Henry single-seat ramjet helicopter. None of them was to enter production, but each brought together a cadre of engineering personnel with invaluable experience, and added to the growing relationship with what would soon become the US Air Force.

The 100-aircraft order for Phantoms was ultimately reduced to 60, but orders for 895 Banshees soon followed. While similar in appearance to the Phantom, the Banshee was a much more sophisticated aircraft. The Phantoms essentially became operational trainers for the next generation of jet fighters, just as the Bell P-59A had been for the Lockheed P-80. The FH-1 entered service aboard USS *Saipan* with VF-17A, which thus became the first shipboard jet fighter squadron in the world. The US Marines began their long association with McDonnell products shortly thereafter, with VMF-122 and VMB-311 receiving Phantoms in 1948.

The Banshee turned out to be about 10,000lb heavier and 100mph faster than its predecessor, with a maximum gross weight in the F2H-3 model of over 25,000lb and a top speed of 580mph. Bob Edholm made the first XF2H-1 flight, on January 11, 1947.

With the Banshee began the formulation of the concept

which would lead to the Phantom II and account for its widespread acceptance and success. The McDonnell engineers turned increasingly towards what was termed "multi-mission engineering," and the Banshee saw itself transformed from a red-hot interceptor and air-superiority type into a fighter-bomber, a night fighter (F2H-2N), a reconnaissance aircraft (F2H-2P) and an all-weather fighter (F2H-3 and F2H-4). The Banshee performed exceptionally well in Korea as an escort fighter, fighter-bomber and reconnaissance aircraft. When faster fighters had supplanted the F2H-3 in US Navy use, 39 examples were transferred to the Royal Canadian Navy, with which they equipped two squadrons aboard HMCS Bonaventure.

The Banshee gave McDonnell substance and a reputation upon which the future could be built. As a result, Banshee project engineer Herman D. Barkey was identified by James McDonnell and current McDonnell president Don Malvern as a man to be reckoned with. He would justify their faith, becoming the unifying force behind the Phantom II, flogging one and all daily to obtain the maximum performance, and rewarding daring and imagination in the solution of the hundreds of problems which would arise.

Even while the first Phantoms and Banshees were laying a course for the Phantom II to follow, McDonnell Aircraft was learning from other less fortunate endeavours. Barkey had proposed the totally unorthodox XP-85 in the autumn of 1944, when he saw a need for an escort fighter to be carried aboard either the Boeing B-29 or the new Consolidated B-36 bomber. The Goblin was an improbable-looking barrel of a fighter, designed to be carried in the bomb bay of the mother aircraft and dropped in flight to provide air defence. Instead of landing gear the XP-85 had a trapeze hook-up system not unlike that fitted to the Curtiss F9C-2 Sparrowhawks which the US Navy operated from the *Akron* and *Macon* dirigibles.

The XP-85 was too hot to handle and too difficult to hook up, and the project was dropped, with the test aircraft winding up in Air Force museums. In a more extended test programme McDonnell continued its long love affair with the Platt-LePage helicopter patents, creating the world's first twin-engine, twin-rotor helicopter, the XHJD-1 Whirlaway. The first flight of the Whirlaway (named after a sensational racehorse of the time) was made on April 27, 1946, by C. R. Wood, and testing continued for the next five years. No orders were secured and the project was abandoned; the Whirlaway is now in storage at the National Air and Space Museum in Washington DC.

Other rotary-wing concepts were tried during the period, including the ramjet-powered Little Henry and the XV-1 convertiplane. Both aircraft succeeded in proving their concepts, but neither looked capable of carrying out its tasks with a full military load. They were both abandoned, ending up like the Whirlaway in the National Air and Space Museum. With the passing of these aircraft McDonnell's infatuation with vertical take-off aircraft had to be satisfied

**Right: An important trio: an F2H Banshee (top), followed by a Demon and an early Phantom II.** *(MDC)*

**Below: Shipmates: the mature Demon and the young Phantom II.**

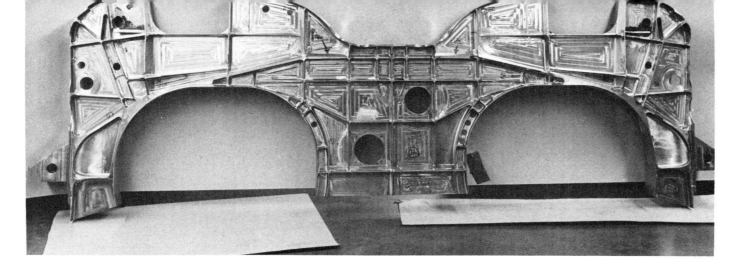

Above: The Phantom II successfully pioneered the use by McDonnell of large forged and machined parts. This technique was used to an even greater degree in the F-15 Eagle. *(MDC)*

with proposals until the advent of the later and much more successful AV-8B Harrier II development of the British V/Stol combat aircraft.

The stage had been set by the Banshee and its contemporaries for two key aircraft in the McDonnell pedigree. The first of these was the sometimes ill-starred F3H Demon, which was ordered for experimental purposes by the US Navy in September, 1949. The second was the F-101 Voodoo, which was ordered by the USAF in January 1952.

The F3H was designed by a team headed by Richard Deegan, and though the airframe was advanced, with its sharply swept wing and tail surfaces, the assignment of the Westinghouse XJ40 turbojet as its powerplant proved to be a curse. The single-engine configuration was foreign to McDonnell, but the Navy insisted, since the original specification for the Westinghouse engine promised enough thrust for the job. The design problem was further compounded when the Navy decided that what had started out as a day fighter should be developed as an all-weather night fighter. Though the initial prototype was first flown on January 7, 1951, by Bob Edholm, the first night fighter did not take to the air until December 24, 1953. The F3H-1N, as it was designated, suffered a series of crashes due largely to the low thrust and unreliability of the Westinghouse engine. Production was stopped after 58 aircraft had been completed, and many were barged ignominiously down the Mississippi to be used for maintenance instruction at Naval Air Station Memphis.

The situation was rectified to a degree when at last McDonnell was able to persuade the Navy to permit the substitution of the Allison J71-A2 engine of 9,500lb thrust (dry) and 14,250lb (wet). A number of modifications were applied to the basic aircraft, including an increase in wing area from 442 to 510ft², and production was resumed.

The Demon finally entered service in March 1956 and served until September 1964. Powered by the Allison engine, the Demon served creditably with 11 Navy squadrons and was well liked by its pilots. But perhaps its most important contribution was to teach McDonnell a number of lessons about dealing with unreasonable military specifications, and about the importance of having a firm say in the choice of engine supplier.

In contrast to the Demon, the Voodoo entered service quickly and was very well received. The Voodoo was especially important to McDonnell because it was the company's first production aircraft for the USAF, and because it embodied more than any previous McDonnell product the multi-mission concept.

Based directly on XF-88 work, the F-101 was created by a design team led by E. M. "Bud" Flesh, who at Curtiss Wright had created the advanced but unstable XP-55 Ascender canard fighter. Flesh's problem was just the reverse of Deegan's, for instead of being saddled with unreliable engines, he had to put the churning power of the magnificent Pratt & Whitney J57 to fighter use. This unit, P & W's first attempt at a large turbojet, played a vital role in US military and civil planemaking over the next ten years, powering the B-52 bomber, the Boeing 707 airliner, the North American F-100 and Convair F-102 fighters, and many other types.

The F-101 made its first flight at Edwards Air Force Base on September 29, 1954, with Bob Little at the controls, and by early 1957 was being delivered to service units. The F-101s set a series of records, with one model following another until 809 had been built. The F-101A was designed for the penetration escort role and could itself carry a large atomic weapon. Next came the RF-101A, a reconnaissance aircraft, and then the F-101C strike fighter, specially strengthened for its low-altitude nuclear delivery role. Canada was impressed with the aircraft, buying both new CF-101Bs and refurbished USAF examples and keeping them in operation from July 1961 until the time of writing. The F-101B interceptor version followed.

The Voodoo taught McDonnell much about supersonic aircraft, from manufacturing techniques to the solution of pitch-up problems and the vexations of increasingly sophisticated electronics. In the process of creating the Voodoo and the aircraft which led to it, McDonnell had established a modern factory and a highly trained workforce, and had won the confidence of the US Navy and the US Air Force. It had experimented in a number of areas, outside the realm of aircraft, ranging from plasma jet research to guided missiles, and had created a financial base which could sustain rapid and continuing expansion. The stage was now set for the Phantom II.

Above: As orders poured in, McDonnell had to expand its facilities. Thrifty Scot though he was, "Old Mac" never economised on tools or equipment, and McDonnell was always ahead of its competitors when it came to factory equipment. *(MDC)*

Below: The Phantom II was built without consideration for some of the manufacturing niceties found in later aircraft. An aphorism of the time stated: "If you can get your hands inside a compartment, something has been left out." *(MDC)*

Below: At the start of the programme everyone in St Louis, including Mr Mac himself, would have laughed at the notion of 5,000 Phantoms being built. The original order was for 23 development aircraft; when production finally ended 5,195 of the husky twin-engined fighters had been built. *(MDC)*

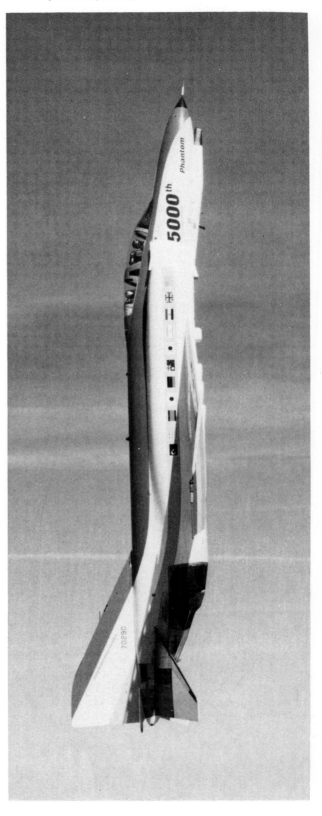

# 3. Designing the Phantom

The XF4H-1 was well on its way to completion when the matter of naming it came up. McDonnell had been doing well in the spirit world, with the success of the Phantom, Banshee, Demon and Voodoo more than offsetting the lack of interest in the Goblin. Don Malvern was then project manager for the XF4H-1, and he wanted to call the big, angular aircraft "Satan". It seemed a fitting name, but when he tried it out on Mr Mac he got a negative response. Worse, he got a suggestion: McDonnell thought "Mithras" would be nice. Malvern blanched: the name of the Persian god of light would be the butt of a million jokes by inventive young fighter pilots.

Malvern was not a manager for nothing, and he set up a mini-project to name the aircraft. He sent out a listing of what some recall to be several hundred names, including Mithras, Satan and Warlock, to potential customers, employees and others with an interest in the project. The replies came in, giving "Satan" as the clear winner. Malvern took the results into McDonnell, who read them carefully and put them aside.

The morning before the christening ceremony the name had still not been selected and the programmes, signs and aircraft were all still lacking the necessary punchy title. Malvern went in to see Mr Mac, hoping to get a decision to use the Satan signs he had already prepared. McDonnell handed him a piece of paper with "Phantom II" written on it. He was the boss, and he was right again.

McDonnell was almost always right, and never more so than in the depressing days of 1953 when it seemed that for the first time since 1945 McDonnell was out of the fighter business, at least temporarily. The Chance Vought XF8U-1, a sleek single-engined fighter with a novel adjustable-incidence wing, had bested the McDonnell submission in a contest for a production contract. The Crusader, as the F8U became known, had a top speed of 1,120mph and was the first Navy fighter to exceed the magic 1,000mph mark. The type gave good and long service, and ultimately 1,308 were

**Below: The F3H-G was a single-seat, long-range attack aircraft powered by the Wright J65 engine of about 10,000lb thrust and armed with four 20mm cannon.** *(MDC)*

Above: The mock-up was adapted to show how the General Electric J79 engine would fit. A few months later, on October 18, 1954, the project was redesignated AH-1 when the Navy issued a letter of intent to buy two of the attack aircraft. *(MDC)*

built. This was a formidable number for a supersonic fighter at the time, and a lesser man than McDonnell might have turned his attention elsewhere. But the indomitable Mr Mac had a dynamic team, and he and David S. Lewis, a future president of the company, had become almost legendary salesmen. He decided to build an aircraft which the Navy would have to buy: his only problem lay in determining what the Navy thought it needed.

In late summer 1953 Herman Barkey, Dave Freeburg and a half a dozen engineers were assigned to an "advanced design cage" to respond to this new challenge. David Lewis was named chief of advanced design, and Barkey was titled project engineer, having served in the same capacity on the Banshee. Frank Laacke was the lead aerodynamicist, under the direction of George Graff. Mike Weeks handled stress analysis, and Gil Monroe was assigned the vitally important task of weight computation and control. Later Bill Blatz became involved in the equally crucial job of designing the variable inlet.

Between October 1953 and November 1954 McDonnell and Lewis made numerous trips to Washington, carrying proposals and closely questioning not only Naval aviators but also their wives about what was needed: Barkey and others had found that the Naval flier would opt for a hotrod single-engined fighter in front of his peers but would confide to his wife his real affection for twin-engined safety.

The initial efforts by Barkey's team resulted in a sort of hybrid Voodoo/Demon. Initially designated F3H-G, it was cleaner in appearance than its predecessors and much cleaner than its successor. Looking back, Barkey believes that these early schemes would almost certainly have resulted in "a pretty mediocre fighter".

McDonnell and Lewis's missionary work bore fruit on October 18, 1954, when the Navy issued a letter of intent to buy two prototype AH-1 attack aircraft. The contract was a boost, but it still left McDonnell with no specific mission requirement, for the Navy Attack Desk was committed to buying Douglas A-4s (and would be for the next two decades) while the Fighter Desk was preoccupied with the Vought F-8. Cdr (later Capt) Francis X. Timmes was project officer for the AH-1 and he worked diligently with the Chief of Naval Operations to define a role for the aircraft. Capt (later Admiral) Noel Gayler, Cdr George Duncan and Timmes visited the McDonnell plant and in a very brief meeting outlined the fleet air defence mission, which required an aircraft which could fly as much as 250 miles out in advance of the fleet, maintain a combat air patrol for two hours, engage in combat, and then return safely to the carrier – all within a three-hour cycle time. Gayler specified that the aircraft should carry missiles only: guns were regarded as a thing of the past.

Barkey and Timmes then began a close working relationship, monitoring the progress of the new General Electric J79 engine, which had been created for the Air Force's Convair B-58 Mach 2 bomber and the equally hot Lockheed F-104. They also evaluated a series of missiles. Finally they chose the J79 in preference to the Curtiss Wright J65, and the Sparrow III missile. Barkey was thus on his way to creating what today would be called an "advanced-technology fighter". The single-seat AH-1 with J65s was expected to have a maximum speed of about Mach 1.5. The J79-powered XF4H-1, as it was designated on May 26, 1955, would have a speed of well over Mach 2, plus an extraordinary load-carrying capability.

In those relatively red tape-free days the engineering team was left pretty much alone, and Barkey and Timmes had carte blanche. As an indication of the simplicity of the times, the all-important decision on whether the new type would have one seat or two was settled when Barkey met Timmes and Duncan on a Friday afternoon after lunch. By 4.30pm there had been a firm decision to go with two seats; internal fuel tankage was reduced by 150gal and a centreline tank

Above: May 26, 1955, saw another change of direction: now the Navy wanted the F4H-1, a carrier-defence missile fighter. The mock-up was completed on November 17, 1955, at about the time that Mr Mac sent a letter to project manager Don Malvern notifying him that "to date the mock-up has cost $637,000". The Sparrow missiles were to be launched from extendable rails. *(MDC)*

Left: Roll-out of the Phantom II was not without its headaches. Mr Mac had not given the final decision on the name, and signs had to be prepared hastily. Then, when the wife of the Assistant Secretary of the Navy, Mrs Cecil Paton Milme, performed the traditional christening rite, the bottle of champagne would not break. Ultimately Mr Mac (foreground, right) popped the cork and poured the bubbly over the nose. *(MDC)*

Below: The first F4H-1, BuNo 142259, on the runway at St Louis. The second crewman is almost submerged in the fuselage but the basic lines of cranked wing and anhedral tail are evident. *(MDC)*

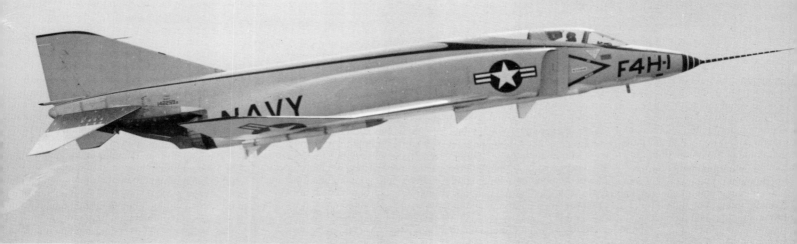

**Above:** No 1 Phantom, fitted with Sparrow mock-ups but almost devoid of markings. *(MDC)*

**Left:** Things went rapidly despite the flurry of changes, and on May 27, 1958, Robert C. Little made the first flight of the F4H-1 at Lambert Field, St Louis. On the left is David S. Lewis, company project manager and later chief of General Dynamics; on the right is Herman D. Barkey, the senior project engineer and the man most people regard as the main thrust behind the programme.

**Below:** No 2 Phantom, 142260, was lost on high-altitude flight, killing the pilot, Zeke Huelsbeck. The Sparrows are semi-submerged in belly indentations, an innovation which resulted in a better than expected drag reduction.

holding 600gal was added. The cockpit profile was altered to provide better visibility for the man in the back seat.

The growth in performance brought about some spectacular changes in configuration, none of them an aesthetic improvement. The original AH-1 had a low-set tailplane, but in the F4H-1 there was nowhere to put it. After much wind-tunnel work and consideration of more than 20 different arrangements it was decided to apply 23° of anhedral. This gave the necessary degree of stability while still leaving the tailplane clear of the jet efflux. This solution, did however contribute, as we shall see later, to severe stall-recovery problems.

Similarly, it became clear that there was not enough dihedral in the new configuration. The centre section of the wing had been planned as an immensely strong unit, spanning 27ft from wing-fold to wing-fold and with a single permanent manufacturing splice at the centreline, and to avoid having to re-engineer the entire wing the outboard panels were simply given 12° of dihedral, imparting an "average" dihedral of 5°. Like so many of the design decisions, this was a commonsense solution which worked.

For an advanced-technology fighter (though the term had not yet been coined) the F3H-G/F4H was born in a time of relatively unsophisticated managerial procedures. The computer and the Xerox machine had still to establish their tyranny over reason, and there were none of the elaborate government regulations calling for weight control, configuration control, maintenance man-hours per flight hour guarantees, fatigue programmes, and so on. The evolution of the F-4 paralleled the introduction of these measures, and was in fact subject to the problems which brought them about. But the McDonnell engineers were not unconscious of the need for concern in all of these matters: it was simply that they were still able to put performance first. The following three short examples, which will be treated at greater length later, illustrate the comparative simplicity of the methods used to produce a highly complex aircraft.

Bill Eaton, a dynamic, burly man, went to Barkey with a suggestion for more access panels to permit easier maintenance. Barkey literally threw him out of his office, for

**Above:** All of the early Phantom IIs were built in the production shops. This is the No 6 aircraft, pictured aboard USS *Independence* during carrier suitability trials, which began on February 15, 1960. *(MDC)*

**Below:** Before being assigned a fleet billet, the first F4H-1 crews and maintenance personnel joined a replacement air group (RAG) for training. On the West Coast VF-121 performed this work at NAS Miramar, with Detachment A of VF-101 doing likewise at NAS Oceana on the East Coast. Here representatives of the two units meet for a symbolic formation flight. The aircraft are later-model F4H-1Fs with revised canopy and radome. They were later redesigated F-4A. *(MDC)*

more panels meant more drag, and Barkey was shooting for performance.

Barkey himself was charged with weight control, along with Gil Monroe, Floyd Stams, Harold Good and George Proctor. Despite their very best efforts the F-4 gained a pound a day, every day for more than 18 years. The reason was not lack of control, but the continuing need for new equipment, new armament, more fuel and so on, and their refusal to make any weight-saving changes that would interfere with production.

Finally, after the first two XF4H-1s had been delivered and the A models were streaming down the line, it was decided to carry out a fatigue-life investigation. To the utter dismay of the McDonnell engineers, the initial fatigue life of the powerful new fighter came out at just 500 flight hours. The armed services were gearing up to make the biggest buy of the most expensive jet fighter in history, and the aircraft would have a potential life span of no more than 20 months! Obviously McDonnell came up with the fixes, for fatigue life now exceeds 5,000hr and Phantoms are programmed to remain in service almost into the 21st century, but in 1958 the outlook was grim.

During the initial design work Barkey adhered to some simple principles which were going to stand the type in good stead over its lifetime. He sought to keep frontal area to a minimum in order to hold down drag, and as a result the first XF4H-1s had a stiletto nose, shielding a 24in radar and a low-profile canopy. He had to keep the length to under 56ft since contemporary aircraft carrier elevators were only 58ft long. A weight budget was allocated to each design area, and no overweight part could be submitted unless another part was reduced in weight by the same amount. There is an unwritten philosophy at McDonnell that all parts should initially be designed at 90 per cent of the required strength so that total weight can be minimised. Then, during the testing process, those parts which prove to be not tough enough are beefed up, resulting in an airframe which is virtually as light as possible for the required strength.

Throughout this period the US Navy was planning to buy two new fighters, from different manufacturers and using different engines. This was sound policy, well proven in the past, for it prevented a sudden grounding order from wiping out an aircraft carrier's defensive capability, as might happen if only one type of fighter was in the inventory.

With the Navy-inspired changes to the XF4H-1 seeming to result in a very high-performance aircraft, Chance Vought was asked to submit a competing single-engined fighter based on the successful Crusader. The response was the single-seat XF8U-3, powered by a Pratt & Whitney J75 and equipped with three Sparrow missiles. Beginning on February 5, 1958, George Spangenburg's lean and efficient Evaluation Division of the US Navy's Bureau of Aeronautics carried out an intensive paper evaluation of the two aircraft, comparing everything from performance to loiter time, from catapult suitability to comparative costs over the lifetime of the programmes. The results were clear: the XF8U-3 was slightly more capable in numerous areas. Its maximum drag-limited speed was slightly higher; it was faster at low altitudes; it was faster in a supersonic climb; it had better acceleration; it required less wind over the deck for take-off or landing; it was more manoeuvrable; it had greater range; and it cost 27 per cent less for an initial quantity of 83 aircraft.

The margin of superiority in each of these areas was small, however, and the Phantom II had some minor performance advantages of its own. What is more, when all the programme costs were considered the advantage fell to as little as four per cent for the Vought design. And the Phantom had two inestimable advantages: two engines and a two-man crew. The two engines conferred a high degree of safety, as demonstrated by the 20 per cent lower loss rate of the twin-engined McDonnell F2H compared with the single-engined Grumman F9F. The two-man crew could make much better use of the outstanding Westinghouse APQ-72 radar that would be fitted, obtaining a "few extra sweeps" on the radar set; three extra sweeps translated into six seconds of time or three miles of distance at high closing speeds.

Spangenburg concluded that the XF8U-3 was a better flying machine, with generally superior performance and lower weight and cost. The F4H-1 had a significant advantage in kill potential and was estimated to be 20 per cent safer and 25 per cent more capable. He recommended that if the money was available the Navy should buy both aircraft; if not, the McDonnell product should get the verdict.

The famed flyoff between the two aircraft at Edwards Air Force Base was in fact not that at all, but a validation of the performance of the two aircraft and a last-ditch effort by the Navy to attract sufficient funds to bring both types into the fleet. As things turned out, the flyoff bore out the

**F4H-1Fs with the large centreline tank and two wing tanks. Early Navy preference was for wing fuel tanks in order to avoid off- and on-loading in the event of engine maintenance.** *(MDC)*

Right: Six Sparrows appeared at first to be a formidable armament but were in fact hardly enough at that stage of the missile's development. This F4H-1F, 146817, was the first to have the larger, 32in, radar dish and revised canopy. *(MDC)*

Below: To demonstrate the F4H-1F's enormous air-to-ground potential this aircraft (BuNo 145310) was loaded with 22 500lb bombs and fitted with a jury-rigged release system using racks, switches and circuitry scavenged from USAF and USN sources. *(MDC)*

Bottom: No 5 F4H-1F was assigned to Navy test squadron VX-5 at El Centro, California, and was flown there by Maj-Gen Hal Vincent (USMC, Retd), who probably has more F-4 time than any other pilot. This was the first Phantom to have boundary-layer control installed. *(MDC)*

Above: Every sort of tactic had to be practised with the Phantom II, including the Navy's probe-and-drogue refuelling. Here a Douglas KA-3B Skywarrior offloads to a VF-101 Det A aircraft. *(MDC)*

predictions pretty closely, though not without monumental efforts on the part of both teams.

Robert C. Little had made the first flight of the XF4H-1 (BuNo 142259) on May 27, 1958; this and the next six flights were carried out from St Louis Airport. There were enormous problems to be solved, and the aircraft was to be at risk during the June-December competition at Edwards. Of all the difficulties, the most severe affected the intake system and the brakes. The variable inlet and the engine bellmouth were simply not working, and the brakes were definitely undersized. But while the brake problem was serious, the intake's malfunctions might have spelled the end for McDonnell's hopes for a Navy order. Ironically, the brakes caused a fatal accident at Edwards when a McDonnell engineer was killed by the explosion of an overheated brake.

The success of the J79 engine depended upon the full development of its thrust potential, which in turn depended upon the air induction systems. The slab-sided air intakes were mounted two inches out from the fuselage to avoid the slow-moving boundary-layer air next to the skin, since slow air means low energy. Each intake system employed a variable inlet ramp, a bypass bellmouth and auxiliary air doors. The ramp was supposed to control the incoming airflow by changing the inlet area and correctly positioning the supersonic shockwave in the duct. The amount of air admitted to the engine itself was controlled by the bypass bellmouth, which opened and closed to maintain the desired mass flow. The excess spilled air was used to cool the engine compartment, and then directed over the afterburner and into the exhaust to increase thrust.

Elegant though it seemed, this system could not at first be made to work. The very complex controls did not govern the variable ramp properly, and engine stalls were severe. The scheduling mechanism, which related the Mach number to the perceived pressure and controlled the bellmouth, was too sensitive and too complicated. The ramp, which was positioned by inputs from the air data computer, had similar problems.

Bill Blatz and a small group of engineers sweated over the problem at Edwards. The cams which operated the air data computer had their profiles altered by hand to suit each day's flight. Don Kingsborough did a similar job, hand-crafting and positioning the bellmouth. Blatz suspected that the curved lower section of the swept-back inlet lip was causing air separation and hence some of the problems. So he simply took a hacksaw and cut eight inches from the lip, put in an aluminium plug, and shaped it with a file.

The small team worked 20 hours a day, seven days a week, in concert with an equally determined General Electric team. Engines were rebuilt overnight, parts were cannibalised from spare engines, and each morning the big F4H was ready to go. Finally, all of these efforts came together in a jury-rigged system which adequately co-ordinated the inlet ramp and the bellmouth, and which was optimised daily for the expected ambient temperatures and pressure altitudes.

On December 17, 1958, the McDonnell product was accepted as the fleet all-weather fighter, and a limited production contract for 24 aircraft was received from the Navy. It was the start of an era that will probably not draw to a close until after the year 2000.

The Vought team put in an equally brilliant effort, solving some inlet problems of their own, but the Navy did not believe that it could obtain enough money to build both types and the XF8U-3 was relegated to test work.

In a single design McDonnell had achieved an aircraft that was outstanding as a missile-carrying fleet defence fighter, an interceptor, a ground-attacker, a reconnaissance aircraft and a record-setter. The following table shows how the new fighter compared with its predecessors, the Demon and the Voodoo. And when it was fully developed it would do even better.

| Aircraft | Max speed | Service ceiling | Initial max rate of climb | Max range |
| --- | --- | --- | --- | --- |
| F3H-2N Demon | 727mph | 42,650ft | 14,350fpm | 1370 miles |
| F-101B Voodoo | 1134mph | 54,800ft | 49,200fpm | 1930 miles |
| F4H-1 Phantom II | 1459mph | 59,400ft | 48,300fpm | 1750 miles |
| F4E Phantom II | 1485mph | 62,250ft | 61,400fpm | 1885 miles |

In the course of its development the Phantom II edged technology onwards in a number of areas. Aerodynamically it combined a Mach 2 dash capability with landing speeds low enough to permit operation from a carrier. This was achieved by fitting the broad, large-area (530ft$^2$) wings with large flaps and a sophisticated boundary-layer control system, introduced on the seventh aircraft. In later models drooping ailerons and leading-edge slats improved low-speed handling still more and cured a problem that limited the type in air combat.

**Above: This head-on shot of a be-bombed Phantom shows how much frontal area was added by external ordnance. Remarkably, performance was little affected, a tribute to the F-4's strength and power-to-weight ratio.** *(MDC)*

Top: The F4H-1 was redesignated F-4B and subjected to the usual experimentation with ordnance. This one carries Mk 83 bombs and LAU-10A rocket pods. Of more interest is the enlarged radome, enclosing the 32in dish for the APQ-72 radar, the 10° fixed and 14° variable inlet ramp, and raised rear seat for the observer. *(MDC)*

Internally there were several innovations. The wiring had become a nightmare: "elephant's trunks" of wires, as thick as a man's thigh, had grown up as one system after another was introduced. McDonnell developed a wire bundle miniaturisation programme that shrank both insulators and conductors down to a quarter of their original size. The structure itself was built up from as many large-formed components as possible, although this technique did not really come into its own until the advent of the F-15.

McDonnell made numerous innovations in metallurgy, using the most advanced titanium manufacturing techniques in the industry. There were improvements in manufacturing methods, with chemical milling, a process by which excess metal is dissolved away in an acid bath, being used on a scale never previously attempted. Large pressings and forgings, while expensive, cut the number of parts, reduced weight and increased strength. Best of all, they lowered stress and vulnerability to corrosion by reducing the number of fasteners. Wing skins were heavy and tapered throughout. Later in the programme composite materials would be tested and then introduced.

The development of such a complex aircraft required an equally sophisticated management to foresee the manufacturing plant required and to obtain the finance needed to procure the presses, forges and machine tools. The initial investment was high but it proved to be prudent, for as orders began to flow in the costs could be amortised over a successively larger number of aircraft. The result was a

**Below: A Phantom from USS *Forrestal* drops its gear. Note the unusual centreline store – a towed-target system – and the dogtooth wing leading edge, incorporated to improve stability** *(MDC)*

Above: A VF-74 Phantom comes aboard the *Saratoga*.

Left: Taken during the working-up period, this photograph shows an early Phantom about to launch, with Demons and a Skywarrior in the background. *(MDC)*

Below left: The Phantom was soon introduced into combat. A VF-41 F-4B is launched for a combat air patrol from USS *Independence*. *(MDC)*

steady reduction in unit cost, which at one point reached a low of $1.7 million.

The history of the growth in size and scope of the Phantom series has been told in several other accounts. What has not been done is to look at how the Phantom matured through a succession of changes to its structure, equipment and utilisation.

The long series of requirements, decisions, compromises, insights and engineering legerdemain which characterised initial Phantom development finally resulted in a two-seat, twin-engined fighter measuring 38ft 4⅞in in span, 58ft 3¾in long and 16ft 3in high, and weighing 28,000lb empty. Fully loaded, the new-born Phantom II weighed 44,600lb.

The initial order for two XF4H-1s had been followed by an award for five pre-production aircraft and then, in December 1958, by an order for 40 F4H-1s. The intended General Electric J79-GE-8 powerplant was not yet available, and the earlier -2 was substituted. The aircraft thus powered were designated F4H-1F, with the last F indicating the lower standard of powerplant. Later the F4H-1F was redesignated F-4A under Secretary of Defence McNamara's tri-service identification system of 1962, while the next series, powered by the -8, was designated F-4B.

The first 18 aircraft had a nose sized for a 24in radar antenna, and the lower canopy line seen on the prototypes. Aircraft Nos 19 to 47 incorporated a 32in radar antenna and a raised canopy. All aircraft from No 48 on were fleet-configured, with the definitive engine, inlet ramp and strengthened landing gear. The last of 47 F-4As was delivered on September 14, 1961, while the last of 696 F-4Bs entered service on January 27, 1967.

The fuselage of the F-4B was built in three main sections: forward, centre and aft. The bulk of the forward fuselage structure was made up of machine extrusions and conven-

Above: The F-4B suited the Marines perfectly. Here is a VFMA-323 ("Death Rattlers") aircraft operating from Da Nang, South Vietnam, during 1966. The Marines received their first F4H-1 in June 1962. *(MDC)*

Right: The moment when all the effort seems to be worthwhile: sunset and Phantoms on USS *Forrestal*, February 1962. *(MDC)*

tional aircraft sheet metal, with some forgings used in high-stress areas. Chemical milling was used extensively to reduce weight. The critical shape of the leading-edge ducts (so adroitly transformed by Blatz at Edwards) had to be repeatable in all aircraft to ensure a happy engine/airframe mating. As a result it was built up from closely spaced ribs and covered by heavy-gauge metal skin.

The 22ft-long centre section incorporates the main bulkheads, which transfer fuselage loads into the front and main spars. This section features double-wall construction to provide forced air cooling to the fuel, which would otherwise be set boiling by the raging heat of the engines.

A strong, rigid but relatively light titanium keel was used between the engines; this simple structure imparts tank-like strength to the entire aircraft. Titanium skins were applied in the engine area and around the aft fuselage, where ceramic-coated titanium shingles were layered to withstand the exhaust heat. More than 900lb of titanium went into each F-4B, an extraordinary amount for the time and an example of the economy-minded James S. McDonnell's willingness to spend money when it yielded results.

The forward fuselage was built in two halves, with equipment installed on both sides. When it came to assembling the whole, a single person in the cockpit could make all the necessary interconnections.

Large forgings and billets machined down to size make up the 27ft-wide wing centre section; this method of construction is costly to tool up for but provides great strength. The entire torque box area of the wing was sealed to provide two fuel tanks of 315gal capacity each. Both the upper and lower torque-box skins were machined down from 2in-thick billets, with integral stiffeners for rib and spar attachment. The main and forward spars were machined from huge forgings each 14ft long and joined in a permanent centre splice. The wide-track landing gear is accommodated aft of the main spar. The folding outer wing is of more conventional construction but still depends upon a thick, tapered skin for strength.

The wing tapers sharply in chord, thickness and thickness ratio. When wind-tunnel studies confirmed the probability of low-speed pitch-up through early stalling of the tip of the swept wings, the outboard wing chord was extended about 10 per cent to create the characteristic dogtooth leading edge.

The Phantom has actually flown with seven different wings in its long life; the first is described above. The second resulted from the modification to include boundary-layer control (BLC) in the leading and trailing-edge flaps to improve low-speed handling and permit lower approach speeds for carrier landings. The BLC was complemented by a greater trailing-edge flap deflection and 16.5° drooping ailerons. These changes generated more changes, for the drooping ailerons required the installation of a slotted stabiliser to offset an induced pitching moment. The third wing version was necessary because the Air Force required wider tyres with a lower "footprint" pressure to improve the safety of operations at high gross weights. The wing was

Right: VF-142 ("Ghostriders") was the first squadron to take the Phantom into combat, flying strikes from USS *Constellation* against shore facilities on August 5, 1964, in response to the Gulf of Tonkin torpedo-boat intrusions. It was the start of a long war. *(MDC)*

Below right: A posed picture, beloved of base photographers but valuable for the details of equipment, ladders and hoses. VFMA-531 was the first USMC F-4 squadron to see combat in South-east Asia. *(MDC)*

locally bulged by 2-3in on both top and bottom surfaces, resulting in only a very slight deterioration in maximum speed.

The fourth wing had the same sort of leading-edge slats as fitted to the F-4E but had the BLC disconnected. The fifth wing requires a little stretch of the imagination: it was the canard surface used on F4H-1 No 266 (Air Force serial number 62-12200), which had been in succession the prototype RF-4C, the YF-4E and the testbed for leading-edge manoeuvring slats (Agile Eagle) before finally becoming the canard-equipped testbed for a fly-by-wire control system. The sixth incorporated the revised leading-edge slats used by the Navy on the F-4J, plus BLC and trailing-edge flaps capable of being lowered to 60°. The seventh was characterised by the substitution of spanwise rather than chordwise blowing for the BLC.

The F-4B was well laden with fuel tanks, having six fuselage bladders of 2,000gal total and two 315gal wing tanks. External fuel was carried in a 600gal centreline tank, which could be supplemented by two 370gal wing tanks if required. These could be refilled in flight by means of a long, hinged retractable probe located on the right-hand side of the cockpit.

Perhaps the single most important piece of equipment to the kill capability of the aircraft was the Westinghouse APQ-72 fire-control radar, which with its 32in antenna imparted a snoutish look to the previously needle-nosed aircraft. It was an essential part of the Airborne Missile Control System, which included the AN/APA-157 radar for Sparrow III missile guidance, the AN/AAA-4 infra-red target seeker, and the missile activation and firing circuits. Other equipment included a 20kVA generator, an ASN-48 inertial navigation system in later aircraft, an ASN-39 navigation computer and an AJB-3 bombing system.

The F-4B received the J79-GE-8 engine, fielding 17,000lb of thrust in afterburner, and also had new intake ramp geometry, with a fixed forward ramp angle of 10° and a variable ramp angle range of 0-14°. The powerful, lightweight J79 had its first test run in 1954 and went on to equip a host of hot aircraft apart from the Phantom, including the Lockheed F-104, the Convair B-58 and the North American A3J. It was the first production Mach 2 engine in history. The J79 had sprightly acceleration characteristics and a relatively low specific fuel consumption of 0.95gal/hr/lb static thrust at Mach 0.9 and 35,000ft. Only 208.69in long, it weighed 3,852lb for a thrust-to-weight ratio of 4.65lb.

Four Sparrow III radar-homing air-to-air missiles are semi-submerged into the fuselage underside, and four AIM-9E Sidewinder infra-red-homing missiles can be carried on inboard wing pylons. Sparrows can be substituted for these weapons if required. The Phantom II carries virtually any Navy store, and is capable of lifting combinations up to 16,000lb, roughly equivalent to four times a B-17's average bomb load.

The USAF somewhat reluctantly jumped on the Phantom bandwagon in 1962, when it selected the F-110A (shortly to become the F-4C under the common designation system) for TAC. The Air Force took its medicine like a man, agreeing to relatively few changes and retaining such Navy features as folding wings and tail hook. USAF features included a dual-control cockpit, different starting (cartridge/pneumatic starter), wider wheels, boom refuelling receptacle and new radar. The USAF went on to acquire the RF-4C, F-4D, F-4E, EF-4E, F-4G and RF-4E. The Navy in its turn followed the F-4B with the RF-4B, F-4G and F-4J (for the Navy and the Marines), and refurbished the F-4B into the F-4N and the F-4J into the F-4S.

Foreign series include aircraft sold or otherwise supplied to Australia, Egypt, Germany, Greece, Israel, Iran, Japan, Korea, Turkey and the United Kingdom. Details of all of these series and foreign variants can be found in Appendix 1.

Top: In an encounter repeated thousands of times over the years by various intruders and interceptors, a VF-114 ("Aardvarks") Phantom shepherds a Tu-16 Badger. The F-4 was operating from USS *Kitty Hawk* in 1963. *(MDC)*

Above: Phantoms, a Vigilante, Skyhawks and Crusaders formate in a loose gaggle which roughly approximates the later Alpha strike force. *(MDC)*

Below: At a time when all the services were under pressure to "commonalise" their equipment, the Air Force evaluated the F4H-1F and the Convair F-106 Delta Dart in the interceptor role. The Phantom had better performance and required less maintenance, and in March 1962 the McDonnell type was selected as the USAF's standard fighter and tactical reconnaissance aircraft, designated F-110A and RF-110A. Here BuNo 149405 carries a spurious Air Force number, 64-9405. *(MDC)*

# 4. Vietnam: the context of combat

Any examination of the Phantom's combat career must begin with its role in America's bitterest and most costly conflict since 1945, the Vietnam War. At all times capable of completely suppressing North Vietnamese air power if they had been given the signal to attack airfields, SAM sites, depots and so on without restriction, the F-4 units were forced instead to operate under absurd policies and rules of engagement which served only to provide both sanctuary and training for their opponents.

American participation in South-east Asia really began in the early 1950s, when US advisers were supplied to the embattled French forces. But the first tangible American presence after the 1954 Geneva agreement which resulted in the partition of the country was the assignment in late 1961 of units of the US Army's Special Forces. Then, in November of that year, the 4400th Combat Crew Training Squadron (Jungle Jim) sent its Detachment 2a (Farm Gate) to South Vietnam, equipping it with vintage North American T-28s and Second World War Douglas B-26s for the training of South Vietnamese pilots. Farm Gate pilots were themselves soon engaged in combat missions, though in the United States their role was written off as a minor effort in a "sub-limited" war, whatever that meant.

Communist subversion was everywhere in South-east Asia at the time, and in April 1962 it seemed that Laos would succumb to the efforts of the Pathet Lao. Joint Task Force 116 was dispatched to Thailand, marking the start of a build-up that would peak at 600 aircraft in 1968.

Despite clear and repeated declarations from the North Vietnamese leaders, most notably the brilliant defence minister Gen Vo Nguyen Giap, American Defence Secretary Robert A. McNamara evolved a theory that North Vietnam could be persuaded to restrain itself if its "provocations" were met with a "graduated response". It was a theory not previously tested in the history of warfare, but seen often in the films of Laurel and Hardy, in which the bumbling pair would patiently and with dignity watch their car being torn apart before matching the outrage by demolishing their adversary's house. In films it was comic; in warfare it was tragic.

The US Joint Chiefs of Staff (JCS) watched the methods by which the Soviet Union and the People's Republic of China poured supplies into North Vietnam, and how these supplies were in turn funnelled through to South Vietnam via lines of communication (LOCs) that ranged from veritable superhighways like the Ho Chi Minh Trail to invisible footpaths underneath the jungle canopy. The JCS proposed that a classic strategic air offensive against 94 important North Vietnamese targets would seal off the ports, destroy the shipping, knock down the bridges, interrupt power supplies, destroy petroleum stores and arms depots, and in general render the North Vietnamese impotent. This move, generally supported by the US Navy, was opposed by Gen Earle G. Wheeler, the Army Chief of Staff, who agreed with McNamara that the war should be won on the ground in South Vietnam. This strategy appealed to the Army by assuring it a rapid expansion and more funds. Thus began the ugly curiosity of an attempt to save a country by bombing it to death.

In August 1964 the North Vietnamese launched a torpedo boat attack against US destroyers in the Gulf of Tonkin, and President Lyndon B. Johnson permitted US Navy aircraft to conduct limited retaliatory strikes against North Vietnamese targets in a "graduated response" to their "provocation". Unfortunately, the North Vietnamese responded in an ungraduated manner, beefing up their MiG fleet and otherwise conducting themselves exactly as they always had, as a tough, shrewd opponent who was determined to win. They stepped up guerrilla rocket and mortar attacks against South Vietnamese airfields, and Vietcong forces began to be openly supported by regular North Vietnamese units.

The North Vietnamese also built up their aircraft defences. In 1964 they disposed of 1,426 anti-aircraft artillery weapons, 22 early-warning radars, and four fire-control radars. By April 1965 they had expanded this to 2,000 flak guns, many of large calibre, two heightfinding radars, 9 AAA control radars, and many SA-2 sites and radars. By 1972, as we shall see, the whole of North Vietnam, as well as major portions of South Vietnam, was covered by fully redundant radar systems and littered with flak batteries and SAMs.

President Johnson, evidently torn between a desire to win a war and the need to placate an increasingly vocal civil opposition to US involvement, took further half measures. He ordered that air attacks be escalated to maintain a steadily increasing pressure, and authorised strikes on targets just north of the Demilitarised Zone (DMZ), which straddled the border between North and South Vietnam. By March 1965 the Flaming Dart campaign had been superseded by Rolling Thunder. (If American military strategy had been as dashing as its system of nomenclature, the war would have been over in 1963.) In Rolling Thunder attacks were authorised up to the 19th parallel and against lines of communication in Laos. Unfortunately, the most vital elements of the North Vietnamese logistics system were well north of the 19th parallel. In a final inversion of normal practice, decisions on the size and frequency of attacks and the choice of targets were reserved for Washington, often being made at Presidential Tuesday Lunches. In essence, the military commander in the field would recommend targets to the 7th Air Force in Saigon, which would go to CINCPAC in Honolulu; CINCPAC would call the Joint Chiefs of Staff in the Pentagon, who would then go up the normal chain of command through the Secretary of Defence to the President and his advisory National Security Council. Approvals (or disapprovals or changes) would come back by the same labyrinthine route, and even the excellent American communication systems could not make up for the delay, loss of context and the inevitable filtering that occurs in any bureaucratic system but is pestilential in the military.

All of the USAF recommendations for a strategic air war against the North Vietnamese had been aimed at avoiding the need for large conventional ground forces. But while McNamara was systematically limiting the utility of the air forces he began at the same time to build up ground forces. The 3,500-man 9th Marine Expeditionary Force went to Da Nang on March 8, 1965. The first McDonnell F-4Bs arrived at Da Nang on April 12, while the first F-4Cs of the 43rd Tactical Fighter Squadron arrived in October, being replaced by the 12th Tactical Fighter Wing at Cam Ranh Bay in November.

But it was the 45th TFS at Ubon Royal Thai Air Force Base that drew first blood for the USAF F-4s. On July 10, 1965, the eager young Phantom pilots set in motion the first of a series of ruses designed to draw the reluctant MiG pilots into combat. Having noted that the MiGs from Phuc Yen airfield characteristically delayed threatening to attack American forces until they estimated that the escorting Phantoms were short on fuel, the 45th had a flight of F-4Cs deliberately arrive over the target area about 15min later than usual. As in the later Operation Bolo, these aircraft flew at altitudes and airspeeds which would suggest on the communist radar screens that they were F-105s. An early radar lock on to enemy aircraft was obtained, but visual identification (an engagement rule) was not established until the tight-turning MiG-17s were only about three miles away, too close for the preferred Sparrow launch. Two separate engagements then broke out.

The MiGs had no choice but to fight, and they turned as the four-aircraft element of the 45th split into two-ship units. Aircraft 1 and 2 turned right; aircraft 3 and 4 turned left. Capts Kenneth E. Holcombe and Arthur C. Clark were in No 3, with Capts Thomas S. Roberts and Ronald C. Anderson in No 4.

The MiGs' noses sparkled with bursts from their lethal 23mm and 37mm cannon. Holcombe and Roberts accelerated and turned in opposite directions, splitting the MiG attack further. A series of wild gyrations followed, with Holcombe using both his high-g turn and vertical capability first to shake off the MiG attacking him and then launch a head-on attack. This could have been fatal, for Holcombe and Clark suffered some cockpit confusion over the armament setting and had to endure another firing pass from the MiG while they got their signals straight. Then, following the now classic energy-manoeuvrability formula, Holcombe, in full afterburner, went into a 10,000ft dive and then high-g barrel-rolled through another MiG gun attack to wind up behind his quarry and within Sidewinder range. Four Sidewinders were fired: three seemed to miss and the fourth had not impacted when the MiG entered a cloud

**Above: The srvices can move swiftly on a good thing: the F-110 procurement decision was reached in November 1961; the first two F4H-1s (alias F-110As) reached Tactical Air Command at Langley AFB in January 1962; the first of 27 F-4Bs borrowed from the Navy arrived at 4453rd Combat Crew Training Wing, MacDill AFB, on February 23; the last was delivered in July 1963; all were returned by June 1964, having accumulated 9,400hr of USAF flight time. Here they are in April 1964, delivery of the USAF's own F-4Cs having begun in November 1963.** *(MDC)*

**Right: At Air Force insistence the F-4Cs had the minimum of changes from F-4B standard. Folding wings were retained, but the USAF boom-and-receptacle air-to-air refuelling system replaced the USN's probe-and-drogue arrangement, and dual controls were fitted in the rear crew position. The rear instrument panel was lowered to improve the back-seater's view, and J79-15 engines were installed.** *(MDC)*

layer. Holcombe and Clark were sick with disappointment until they heard that the No 2 F-4 had observed their target blow up.

Roberts and Anderson in the No 4 aircraft did not have quite the same tussle with their MiG-17. They made a textbook dive-and-climb attack which positioned them behind the enemy, and got off three Sidewinders. The first two missed but the third blew up just behind the MiG, which went vertically into the ground. It was a good beginning to what was to be a long war.

The first Phantom loss occurred on July 24, 1965, when an SA-2 missile shot down an F-4C and damaged three other aircraft in the flight. The result was a decision to permit attacks only on SAM sites which were actually firing, except of course those located above the 20th parallel.

The North Vietnamese continued to increase the strength and quality of their air force and anti-aircraft and missile defences. On August 24 and 25 seven American aircraft and a reconnaissance drone were shot down. The Joint Chiefs of Staff predictably requested permission for a strike against North Vietnamese airfields. Secretary McNamara refused, saying that to do so might cause China to undertake the air defence of North Vietnam. McNamara persisted in the view that the Vietcong represented a civil insurrection and were somehow unrelated to events in the North. He hoped that if the Vietcong could be suppressed by ground action, North and South Vietnam would then co-exist. He had in fact chosen to forswear the only winning option available to the United States, strategic air warfare, and instead insisted upon presenting the combat initiative to North Vietnam. Later, when his tactics had failed, he recommended withdrawal, knowing then that no co-existence would be possible.

The murderous flak over North Vietnam had already forced a change in US tactics. At first USAF aircraft had attacked with conventional bombs in flight profiles similar to that developed for nuclear weapon attacks. They would come in low, pop up to altitude for a diving attack, and then pull out to a low-level egress. The hot anti-aircraft artillery had forced a revision to more conventional dive-bombing techniques, but then the introduction of SAMs forced the aircraft to come in low again. But not for long, for the dense low-altitude flak was far more dangerous than the SAMs, and it proved more "cost-effective" to come in at 6,000-9,000ft, avoiding the flak and depending upon electronic countermeasures and evasive manoeuvres to defeat the SAMs.

Then, in mid-1965, the North Vietnamese brought in their first MiG-21s, increasing their air-to-air-capability. Throughout the conflict the North Vietnamese Air Force followed a policy of gradually increasing its efforts as strength and proficiency grew, only to draw back sharply and stand down if the Americans scored a significant series of victories. By the latter part of 1966 MiGs had become hard to find, their tactics being designed not so much to bring the Americans to battle as to make the US strike forces jettison their bombs by making an attack seem imminent.

The tough, aggressive F-4 crews continued to seek combat, shooting down 13 MiGs between April and September. The first MiG-21 kill went to Maj Paul J. Gilmore of the 35th TFS, 480th TFW, with 1st Lt William T. Smith as his "guy in back" (GIB). Escorting a formation of Douglas EB-66s, they were attacked by MiG-21s coming in high. The EB-66s broke off and left the area, and the MiGs apparently thought their mission was over, returning in an afterburner climb to their "perch". Gilmore pursued one of the MiGs, which was unaccountably making flying-school gentle clearing turns as he edged past 30,000ft. Gilmore's aim was accurate, putting a Sidewinder into the MiG-21. He failed to see the pilot eject, however, and fired two more Sidewinders at the now pilotless MiG. One of the Sidewinders missed but the other went up the tailpipe, twice killing the MiG.

On July 14 the USAF scored a double when a flight of 480th TFS, 35th TFW, F-4Cs engaged MiG-21s. Capt William J. Swendner and 1st Lt Duane A. Buttell Jr were in the lead Phantom, with 1st Lts Ronald G. Martin and Richard N. Kreips as No 2. They were providing the newly required MiG cover for an Iron Hand flight of radar-busting, Shrike missile-equipped F-105s. The MiG-21s were smoking in on the Thuds from their favourite 7 o'clock position when Swendner and Martin closed in. Swendner sent a Sidewinder over the canopy of one MiG, causing the North Vietnamese pilot to break off the attack by hitting the afterburner and climbing. He was fast enough to avoid the second Sidewinder but not the third, which impacted in his tailpipe. Martin made short work of the second MiG, inserting a Sidewinder in his tail.

The North Vietnamese continued to lie low throughout the rest of the year, and the F-4s were able to claim only two more victories. Much seemed to depend upon the individual MiG pilot, as on November 5, 1966, when an attack was made on an EB-66 doing electronic countermeasures work between Hanoi and Haiphong. Maj James E. Tuck of the 480th TFS swung in behind the MiG-21 in Opal 01; 1st Lt John J. Rabeni Jr was in the back seat. Tuck launched three Sparrows: two missed but the third snuffed the MiG and the pilot bailed out. The second MiG, with a tenacity not often found, attempted to be the fourth aircraft in the aerial daisy-chain but 1st Lts Wilbur J. Latham Jr and Klaus J. Klause in Opal 2 turned in behind it to fire. Latham zeroed in on the MiG and fired a Sidewinder which did its heat-seeking well up the NVAF fighter's tailpipe. Splash two.

Despite their losses the North Vietnamese had reason to be satisfied with their efforts. When they were able either to attack or feint an attack at the F-105s the latter were obliged to jettison their bombs and initiate defensive tactics. The MiGs had succeeded all too well at this, and with their main airfields given sanctuary by the self-imposed US rules of engagement, they were immune to attack in all but a limited set of circumstances.

Every USAF airman, from the Chief of Staff down to the greenest new guy, knew what needed to be done: a surgical strike at the airfields, destroying the MiGs on the ground. With this ruled out, it was decided to lure them into the air with Operation Bolo (see Appendix 2 for the official secret account of this mission).

The legendary Col (later Brig-Gen) Robin Olds, 8th TFW (Wolfpack) commander, was called upon to make a maximum effort which would again have F-4s simulating an F-105 attack. This time the plan called for a complete simulation of the routes, callsigns, refuelling areas, speeds and altitudes which would normally be used by the F-105s. F-4 units from the 8th, 355th, 366th and 388th TFWs were called into play, with the danger-loving F-105s from the 355th and 388th assigned to their regular Iron Hand anti-missile site sorties.

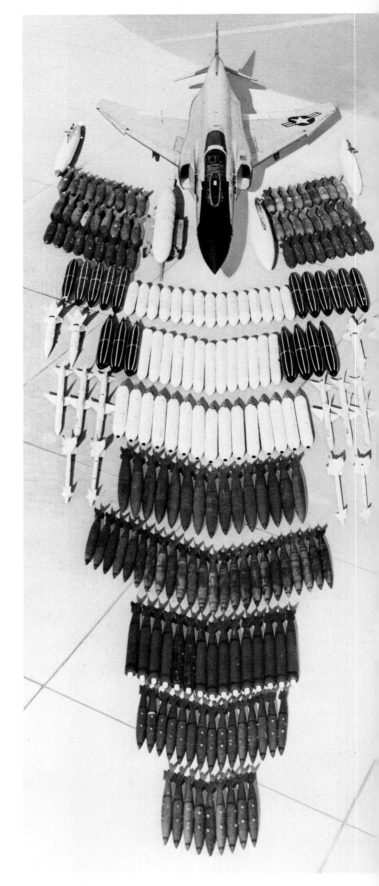

**Left:** One attraction of the Phantom was its immense stores capability. *(MDC)*

**Above:** The F-4C was the first USAF Phantom to obtain a MiG kill. Capts Thomas S. Roberts (aircraft commander) and Ronald C. Anderson (pilot) of the 45th TFS knocked down a MiG-17 on July 10, 1965. This is the F-4C that Brig-Gen Robin Olds used for two MiG kills; the picture was taken later in the aircraft's career, after it had been transferred to a reserve unit. *(MDC)*

The strike went off on January 2, 1967, although weather conditions were bad, particularly over the target area, which included co-ordinates at which the MiGs usually orbited and the Gia Lam and Phuc Yen air bases. The first F-4s of the 8th TFW, Olds Flight,* were led by Olds himself and arrived over the target at 1500hr. Subsequent flights arrived at five-minute intervals. Ford Flight, led by Wolfpack deputy commander Col Chappie James, was second over the target. James, a friend of Olds for 20 years, would go on to become the USAF's first black four-star general.

Predictably, the MiG response was slow and it appeared at first that the entire effort might prove to be just an expensive charade. (Fighter aircraft sorties averaged about $8,000 apiece, and more than 100 aircraft, including the more expensive EB-66 and KC-135 support aircraft, were involved.) Then, with the suddenness that has characterised air warfare since Boelcke's time, there developed a gruelling 15min battle with aggressive MiG-21 pilots, all within a 15-mile radius of Phuc Yen. The MiGs, under strict ground control despite the extent of the mêlée, would attack in two pairs, one from 6 o'clock and the other from about 12 o'clock. The F-4s had to turn to face their nearest attackers, leaving the other MiGs a chance to streak in and launch. Olds, a 24½-victory ace from the Second World War and absolutely adored by the younger fighter pilots of his Wolfpack, gave the following account of his kill:

*US car company names were used as callsigns. It was no accident, and not much security, to have Olds leading Olds flight.

"At the onset of this battle the MiGs popped up out of the clouds. Unfortunately, the first one to pop through came up at my 6 o'clock position. . . . This one was just lucky. He was called out by the second flight that had entered the area. . . . I broke left, turning just hard enough to throw off his deflection, waiting for my three and four men to come in on him. At the same time I saw another MiG pop out of the clouds in a wide turn about my 11 o'clock position, a mile and a half away. I went after him and ignored the one behind me. I fired missiles at him just after he disappeared back into the clouds.

"I'd seen another MiG pop out in my 10 o'clock position, going from my right to left; in other words, just across the circle from me. When the first MiG I fired at disappeared, I slammed full afterburner and pulled in hard to gain position on this second MiG. I pulled the nose up high, about 45°, inside his circle. Mind you, he was turning around to the left, so I pulled the nose up high and rolled to the right. This is known as a vector roll. I got up on top of him and, half upside down, hung there and waited for him to complete more of his turn, and timed it so that as I continued to roll down behind him I'd be about 20° angle off and 4,500 to 5,000ft behind him. That's exactly what happened. Frankly, I am not sure that he ever saw me. When I got down low and behind he was outlined by the sun against a brilliant blue sky. I let him have two Sidewinders, one of which hit and blew his right wing off."

Six more MiGs were shot down in combats basically similar to Olds'. Olds' wingman, Olds 2 (1st Lts Ralph F. Wetterhahn and Jerry K. Sharp), drew first blood by putting an AIM-7E Sparrow through the stabiliser of a MiG-21, knocking pieces off and sending the aircraft into a flat spin. Capt Walter S. Radeker III and 1st Lt James E. Murray III in Olds 04 knocked down the second MiG with a high-speed yo-yo and a Sidewinder in the tailpipe.

Chappie James' wingman, Capt Everett T. Raspberry in Ford 02 (with 1st Lt Robert W. Western as GIB), inserted a Sidewinder into yet another MiG's tailpipe, causing it to swap ends and spin in. Maj Phillip P. Combies and 1st Lt

Lee R. Dutton in Rambler 04 launched two Sparrows at about one mile range and killed a MiG-21 that had been moving in relatively gentle, less than 4g turns throughout the combat. Rambler Flight leader Capt John B. Stone, with 1st Lt Clifton P. Dunnegan Jr in the back seat, came up against two MiGs which had been making persistent cannon passes. Stone initially launched one Sparrow, then salvoed two more. The second Sparrow caught the MiG at the wing root, starting a fire. The pilot punched out.

Two young first lieutenants, Lawrence J. Glynn and Lawrence E. "Larry" Carry, in Rambler 02 blew up a MiG-21 with a Sparrow but collected some incidental damage when they flew through the debris.

The loss of seven MiG-21s in a single fight was complemented on January 6, when Triple Nickle (555th TFS) crews nailed two more and provoked another of the North Vietnamese Air Force's customary post-defeat standdowns. When they emerged again they would forswear the tactics of operating in relatively large units and instead fly in one or two elements of just two aircraft each. Ground control would vector them well outside the visual or airborne radar range of the Phantoms (which were themselves highly visible as a result of their unmistakable radar signature and long black smoke trails), then bring them around to a 6 o'clock position. The MiG-21s would then go supersonic, gathering plenty of "smash" by reaching Mach 1.4 or better, and launch heat-seeking Atoll missiles. After launch the MiG-21s would zoom-climb away to safety.

The tight-turning MiG-17s were usually detailed to defend the approaches to airfields (there were nine bases with runways of over 6,000ft, and perhaps two dozen others with shorter strips), but they would occasionally be vectored in low to strike at the attacking force from below, hoping to break off sections and to cause bomb loads to be jettisoned. If the strike force floundered as a result, MiG-21s would drop from above to attack straggling units.

The artificial sanctuary accorded to the MiGs was removed in April 1967, when attacks on Kep and Hoa Luc airfields were approved. Claims for as many as 32 MiGs destroyed on the ground were made, and the Communists were literally driven back up into the air to oppose the F-4s. It was characteristic of the air war in Vietnam that the United States was successful when it resorted to tactics previously proven in the Second World War or Korea. Neutralising the enemy's airfields has been primary strategy since Billy Bishop's forays in the First World War, and it is today the principal concern of air commanders on both sides in Europe.

For the rest of 1967 the attacks on North Vietnam proceeded at an increasing tempo, with key military airfields, rail yards and power stations the preferred targets. USAF air attacks went from something like 130,000 sorties in 1966 to 176,000 in 1967 before leaping to 221,755 in 1968, a reflection of the urgency with which ordnance had been developed, men trained, aircraft procured and bases made ready. Then the slide began: sorties dropped to 170,000 in 1969, to 81,000 in 1970 and to less than 19,000 in 1971.

Though the anti-MiG war always took the headlines, the interdiction of movement along the railways, roads and trails was the Americans' main preoccupation. With Russian and Chinese help, plus their own inimitable industry, the North Vietnamese were capable in 1968 of generating about 62,000 tons of supplies to send down the Ho Chi Minh Trail. In that same time, US air power reduced the deliveries to Vietcong troops to about 13,000 tons. The North Vietnamese increased their efforts to about 68,000 tons in 1969, and the lower USAF sortie total meant that the quantity delivered went up to 21,000 tons. In 1970, when sorties were down even further, the North Vietnamese made an additional effort and shoved out 68,500 tons. But

**Below: F-4Cs of the 557th TFS at Cam Ranh Bay in January 1968. These aircraft gave long and hard service under very severe weather conditions.** *(Walter House)*

**Left:** *Shehasta*, the F-4C flown by the commander of the 12th TFS in December 1968. *(Walter House)*

**Below left:** *Sharkbait One*, mount of the 557th TFS commander. *(Walter House)*

improved methods, particularly the AC-130 gunships, reduced this flow to less than 10,000 tons.

In the air, F-4s continued to seek out MiGs. The aircraft were newer and better equipped: in May 1967 F-4s began to be fitted with the SUU-16 gun pod, an external store with a rapid-fire 20mm Gatling gun. It wasn't the perfect answer, for it lacked the rigidity of a fixed mount and the ease of boresighting of an internally mounted gun, but it was a big help and had long been called for. On May 14 two MiG-17s were brought down by 366th TFW crews using the SUU-16, and another fell to an AIM-7. All three victorious crews were from the 480th TFS, acting as MiGCAP for F-105s attacking the Ha Dong army barracks. For a change the MiG-17s were up in force in two flights, one of 16 aircraft and one of 10.

Maj James A. Hargrove, in Speedo 01 with back-seater 1st Lt Stephen H. DeMuth, engaged in a long five-minute battle with all sorts of MiGs. Hargrove fired off all of his Sidewinders and Sparrows, and every one missed. By the time his fourth MiG opportunity came along he was more than a little frustrated. Pulling to within 2,000ft of the enemy in a descending right-hand turn, he began firing and closed up to 300ft separation before the MiG-17 flamed and exploded.

The second SUU-16 victory came when Capt James T. Craig Jr and 1st Lt James J. Talley in Speedo 03 had a similar run of good and bad luck. They missed the first two MiGs with their Sparrows but a 2½sec burst from their 20mm was enough to send a third MiG spinning straight into the ground.

The success of the SUU-16 vindicated many of the Phantom pilots, who had long complained that a gun was necessary at short range.

During 1967 the Phantoms were credited with 36½ MiGs, while the F-105s came up with a surprising 22½ victories. Considering the late arrival of the gun installation on the F-4, it did pretty well. During 1967 the F-4s shot down 15 MiGs with AIM-7s, 11 with AIM-9s, 7½ with the 20mm gun, two with AIM-4s, and one by flying it into the ground. The F-105s disposed of 19½ with 20mm fire and three with AIM-9s.

The Tet offensive, which erupted in January 1968, was a tactical loss but a heartening political victory for the North Vietnamese. The proponents of "sub-limited war," "graduated response" and the other concepts which shackled air power now came to the conclusion that the war was not winnable. On March 31, 1968, President Johnson, watching his own hopes for presidential immortality sliding down the Vietnamese drain, announced the cessation of all attacks north of the 19th parallel in an attempt to "de-escalate" the conflict. On November 1 of that year, three days before the presidential election, he extended the bombing halt to all of North Vietnam.

As a result, 1968 was both a slow year for MiG-killing, and

Right: Maj Robert D. Russ and 1st Lt Douglas M. Nelson made a belly landing in this F-4C at Cam Ranh Bay in June 1968. The aircraft was only lightly damaged. *(USAF)*

Below: Another forced landing, this time with one gear leg locked up. Normal recommended procedure in such cases is to bail out, but the pilot elected to try to save the aircraft. The incident took place at Cam Ranh Bay on February 14, 1969. *(USAF)*

the last until 1972. The 8th TFW shot down seven MiGs and the 432nd TRW one, for a total of eight for the year. This does not mean that the Phantom pilots were idle, however, for combat against MiGs was only one facet of the air war. The Phantoms had long been engaged daily in the dangerous task of close air support, and nightly in the even more dangerous business of interdicting supplies in North Vietnam and Laos.

President Richard M. Nixon took office on January 4, 1969, inheriting a war that had gone adrift in a sea of domestic unrest. Buzzwords were now more important than bombs, and the next one that cropped up was "Vietnamisation," or the training and equipping of South Vietnamese forces so that they could defend themselves, permitting the United States to disengage. The North Vietnamese responded with consistency and purpose, launching a massive build-up of troops, radar sites and SAM installations, and replenishing their air force.

The period from 1968 to 1972 refutes the notion that air combat means dogfighting and nothing more. There are all kinds of combat, many of them far less glamorous than bouncing a MiG but all equally dangerous and some requiring a totally different type of courage and skill. The multiple capabilities of the F-4 meant that it could do almost any mission. Since many of them were conducted at night or in foul weather, well out of the range of American ground control, they made tremendous demands of the Phantom crews and their equipment. Standard F-4 missions of the period included Spectre escort, ResCAP, Fast FAC, Night Owl FAC, laser-guided bomb designation and delivery, sensor delivery, mining, armed recce, B-52 escort and MiGCAP. A few of these tasks are described in the following pages.

### Spectre escort

Spectre was the collective callsign for the highly efficient Lockheed AC-130 gunships that did so much damage to supply lines through Laos. The F-4s' job was to suppress the increasingly intense and accurate flak that assailed the gunships. Three F-4s were usually launched so that one was covering the AC-130 at all times. The AC-130 had a typical on-station time of three hours, and all of the missions took place at night.

Each F-4 would usually be configured with two 370gal wing tanks, four CBU-58 cluster bombs and four M36E2 fragmentation bombs. Alternatively, it might have a single 600gal centreline tank, six CBU-58s and two BLU-27s filled with Napalm B.

The flight leader would brief with the gunship commander to discuss the mission and agree on targets and tactics. The leader would then brief the other F-4 jocks. The first aircraft, usually the flight leader but not always, would launch so as to arrive in the target area just after the AC-130. The second F-4 would take off, fly to a tanker orbit area, refuel and then proceed to the target area, arriving just before the first fighter reached bingo fuel. (Bingo fuel was the amount required for the fighter to fly to a tanker orbit and, if unable to refuel, still be able to divert to a recovery airfield.) The first fighter would refuel and then return to the gunship. The third fighter would launch direct to the target area, relieving the second, who would refuel again. Each fighter would fly two periods escorting the gunship; if the Phantoms had enough ordnance and the tanker sufficient available fuel, this could be stretched to three periods.

The gunship would fly along lines of communication in the operational area, and after locating a target would establish a gun-firing orbit around it. The escorting F-4 would fly above and outside the orbit of the gunship and attack any AAA site that opened fire. Visibility over Laos was often bad, the result of weather or smoke from fields being burned off, or a combination of the two. It took courage of a high order to pull a 15-ton fighter into a screaming dive into a pitch-black hole, aiming only at the flash of lights from flak guns. The pilot prayed that his altimeter was right, his maps were right, that he wouldn't be hit, that his back-seater would monitor the pull-out and that he hit whatever was shooting at the gunships.

Capt (now Lt-Col) Richard S. "Steve" Ritchie, the USAF's first ace of the Vietnam War, tells a story about going out as an instructor pilot on a Spectre mission with a brand-new pilot in the back seat. The night was pitch-black and they were having serious trouble locating the target, but the time came when the flak got so bad that Ritchie had no choice but to roll the big beast of a Phantom on its back and plunge down. The flak had disappeared but Ritchie dropped ordnance and pulled out, with the back-seater oblivious of position, attitude, altitude and opposition. A few moments later there was a tremendous explosion, a brilliant flash of light and a series of secondary explosions. The new pilot, still naturally clueless, received an entirely erroneous impression of how easy the work was and flew back with Ritchie to receive a Distinguished Flying Cross for his work.

The danger from the AAA was not as great for the F-4s as it was for the Ubon-based AC-130s, but the job remained extremely demanding and eventually some wings assigned all of the Spectre escort missions to the same squadron so that a higher level of expertise could be acquired.

The mission of escorting the AC-119 Shadow gunships was basically the same as that described above except that F-4s and gunship often did not enjoy the same degree of communication.

### Fast FAC

The Fast FAC (forward air controller) mission originated in 1967 when enemy flak became too intense for the Cessna O-1s and North American-Rockwell OV-10s to survive. Two-seat North American F-100F Super Sabres were used for the first Fast FACs, code-named Misty. The F-100 was then superseded by the F-4. Crews from the 12th TFW at Cam Ranh Bay participated in a series of tests with a qualified Fast FAC in the rear seat. The first F-4 unit to field a Fast FAC was the 366th TFW at Da Nang in the summer of 1968. Front-seat Phantom pilots would be sent for five trips in the rear seat of a Misty F-100, then the Misty back-seat FACs would come to Da Nang for three rides in the F-4.

The purpose of Fast FAC was visual reconnaissance in high-threat areas of Laos, Cambodia and North Vietnam. Missions were flown primarily during the day, with some Night Owl sorties at night. For a daylight Fast FAC mission the F-4 would be equipped with two 370gal wing tanks, two

**Not everyone was so lucky: this RF-4C crashed and burned on April 30, 1969.** *(USAF)*

LAU-32 rocket launchers (each containing seven rockets with white phosphorus warheads) and one SUU-16 or SUU-23 20mm gun pod.

These one-ship sorties were flown by volunteers from amongst the most experienced pilots in the wing. They had to be in outstanding physical condition, able to take up to four or five hours of constant high-g turns at low level. (Ritchie, who had flown as a Misty FAC, flew the first Fast FAC sortie in Thailand and subsequently completed 95 of these hazardous missions.) Back-seaters, either pilots or weapon systems officers, were also volunteers and usually highly experienced combat veterans.

The crews were normally assigned only to Fast FAC work and had dedicated intelligence and administrative support. Briefings were intensive study periods in which the crews reviewed the latest intelligence on the areas they were scheduled to cover, and then created their own mission plans. Each crew would spend three 40-50min periods per mission in the target area, depending upon fuel usage and the location of the tankers.

When a target was found the FAC would request a strike flight from the airborne battlefield command and control aircraft (ABCC) responsible for the area. If there were flights airborne the ABCC (usually a C-130) would direct them to the Fast FAC's target. On rare occasions the ABCC would scramble an alert flight. The strike flight would then rendezvous with the FAC as close to the target as possible. The FAC would mark the target with a white phosphorus rocket or give pinpoint locations by reference to obvious geographical features. The strike would go in and the FAC would evaluate it, calling in another if necessary.

The missions were far from routine: the FAC had to fly at very low altitudes to locate camouflaged targets in the dense jungle foliage, and consequently was very vulnerable. Losses on Fast FAC operations were significantly higher than for standard strike or reconnaissance work – and they never got to shoot at a MiG. But those who flew Fast FAC loved the mission. Not only was it highly satisfying, but it proved to be one of the most successful operations of the war.

**Night Owl**

The 497th TFS, 8th TFW, based at Ubon, was the only dedicated night fighter squadron in South-east Asia. Other squadrons flew at night, of course, but none flew in the dark for month after month as the Night Owls did. The 497th's brief ran from 1800hr to 0600hr, covering dusk, night and dawn, the worst times for close attack work.

**Left: The indispensable combination: Young Tiger KC-135 refuels a Phantom. Some missions required as many as three refuellings, each one progressively more demanding as the pilot tired. Against regulations, many aircraft commanders taught the back-seater to refuel, in order to insure against incapacitation and to obtain a little rest. Once considered an unorthodox, risky and essentially self-defeating tactic, refuelling became so routine as to be boring; many rendezvous were conducted in complete radio silence. Oddly enough, there were very few attacks on the tankers, even though their brave aircrews often entered danger areas to refuel battle-damaged aircraft.** *(Capt Jim Barr)*

Right: RF-4C 65-0903 of the 12th TRS, 460th TRW, based at Tan Son Nhut, South Vietnam. This photograph was taken at Phu Cat on June 27, 1971, following a mission over North Vietnam. *(Norman E. Taylor)*

Below: Just as aerial refuelling became routine, so did the delivery of ordnance – except for those last few minutes of the roll-in when the aircraft became vulnerable to AAA and small-arms fire. Attacks were made from surprisingly low altitudes, and accuracy often required dives to very low levels. Here an F-4C pulls contrails as it turns in. The aircraft retains its Sparrow armament, and could climb to altitude to act as a combat air patrol after dropping the ordnance. *(MDC)*

Bottom: The Phantom was adapted readily to the photo-reconnaissance role, and is still one of the world's best aircraft for this task. Here the faithful old ex-Navy F-4B 12200 has been rebuilt as the prototype RF-4C. This aircraft would later serve as the F-4E prototype, and then test the new system of slats. Still later it flew as the canard-equipped, fly-by-wire F-4CCV prototype, and finally ended up as the testbed for the Precision Aircraft Control Technology programme. *(MDC)*

Upgrading to Night Owl work was not easy: only experienced crews were selected, and even these needed a 15-ride theatre check-out. They had to learn refuelling, gunship escort, flare techniques, visual and Loran bombing, and more. Careful training resulted in relatively few losses despite the hazards of the mission.

Night Owl aircraft were prepared specifically for their nocturnal hunting ground. Cockpit illumination was kept as low as possible, and most internal and external lights were dimmed or taped over. The engine fire-warning lights, located so close together that it was sometimes impossible to tell which was illuminated, were a particular problem. The 497th's F-4s were eventually (after 1969) equipped with Loran and external strip telepanel lighting and their bellies painted black. The Loran set, in combination with Task Force Alpha*, made accurate attacks possible.

The Night Owls had a variety of missions, including Spectre escort, night strikes, FAC, blind bombing, day strikes up North, and search and rescue work. Spectre was the most demanding and dangerous: Night Owls particularly hated the thought of diving through the AC-130's orbit, with the ever-present possibility of a mid-air collision. Worse still for the F-4 pilot was the frustration of having to watch the AC-130s blow up truck after truck while he orbited, waiting.

Night strikes were much like day strikes, except that the FACs would illuminate the target with a rocket or a flare for orientation. Two markers were better than one, for the distance between the two (whatever it was) could be used as the basic reference unit for correcting bomb drops.

When the monsoon brought an end to visual work, night bombing was done either on radar cue or by the Loran navigation system. The far more accurate Loran was preferred. Working with Task Force Alpha, the pilot would "fly an ILS [Instrument landing system]" at altitude to get accurate steering; the drop was then automatic. Task Force Alpha would adjust the drop point by changing the release timing in response to data from sensors placed earlier by low-flying F-4s.

The Night Owl FAC mission was the most coveted and demanded the most training and skill. The dusk patrol was preferred, since at that time the enemy trucks were normally stacked up in North Vietnam, waiting to start. When the sun went down the trucks would run through Ban Karai or Mu Gia passes, and then the Owls would be on them. Loran helped the Owl FAC work, since known points could be pre-set and then checked for traffic. The Owl Phantom crew would fly to the checkpoint, pop a flare and immediately start a climbing turn to be in position to deliver ordnance visually. There were a number of complications: the switches had to be changed rapidly from the settings for dropping flares to those for cluster bombs; the use of afterburner would alert every gunner in the area; and flare-drop airspeed was relatively low, leaving little margin for pull-up or banking and execution of a proper dive-bombing pass. On top of all this, the F-4 had to remain above the flare altitude to avoid being seen by enemy AAA gunners. If the F-4 missed on the bomb run it was difficult to get around again before the flare went out, and then the whole "switchology" process had to be run through again to dispense another flare.

The average mission comprised three trips to the target area, with two refuellings in the process. An ever-present danger was target fixation, especially when the flare ignited and the crew could see a line of trucks going down the road.

The back-seater was regarded as the key to successful night flying. The "pitter" would normally "stay inside" the aircraft, focusing his attention on the cockpit, checking the Loran, altitude, airspeed and position, and in general keeping the aircraft commander (AC) out of trouble. The AC "stayed outside" as much as possible, looking for targets and threats. If they had to change roles for some reason, the AC would call "coming inside" and the back-seater would take on the outside-looking task. Fortunately, the F-4 was a forgiving aircraft once the notorious switchology was mastered.

*Task Force Alpha was an infiltration surveillance centre using information obtained from Igloo White sensors (see below).

Above: The reconnaissance nose was 33in longer and more tapered. The APQ-99 radar is mounted forward, followed by the KS-87 framing camera, the KA-56 panoramic camera, the LA-313 aerial mapping and reconnaissance viewfinder, and the UPD-6 radar mapping antenna. *(MDC)*

### Sensor delivery and Igloo White

In his apparent desire to outspend but not offend the North Vietnamese, Secretary of Defence McNamara approved the idea of the electronic "fence" which would stretch across the hostile countryside. Highly sensitive air-delivered sensors of two types would be placed in areas associated with North Vietnamese resupply activity, particularly in Laos. One of these devices was acoustic, similar in concept to the sonobuoy used in anti-submarine warfare. The other was a seismic sensor which detected vehicular road traffic through ground vibration. The acoustic sensor was employed in heavy jungle areas, where it would remain suspended by its small retarding parachute in the dense foliage. The seismic unit, which was implanted in the ground with only the antenna visible, was used in more open areas. The antenna looked like a tree branch and was not readily identifiable.

Once implanted, the sensors relayed continuous coded data on enemy vehicle and troop activity back to a computerised facility at Nakhon Phanom, Thailand. Specially modified RC-121s orbiting over Laos were used as in-flight communications relays between the ground sensors and the facility at Nakhon Phanom, known as Task Force Alpha. This unit's computers would analyse the data and select for subsequent attack the most active route segments and hidden truck parks along the hundreds of miles of Ho Chi Minh Trail.

The Igloo White sensors had limited detection ranges, and extremely accurate delivery was required if they were to be effective. This meant dropping them from very low altitude, with high speed also being necessary to avoid excessive losses to heavy automatic weapons fire.

The 25th Tactical Fighter Squadron, based at Elgin Air Force Base, Florida, was selected for the mission. After modification of their aircraft and training, the unit deployed to Ubon in May 1968. The 25th also specialised in the delivery of a variety of anti-personnel/vehicular munitions which had been specially developed for the mission.

The 25th was equipped with F-4D-32s, which had been modified to include the extremely accurate Loran-D navigation system. This system, integrated into the Weapons Release Computer System (WRCS), permitted automatic sensor or weapon deliveries, with steering information being provided by the azimuth steering needle on the Attitude Direction Indicator (ADI). Target co-ordinates, sensor/weapon ballistic data, and other tactical information (such as sensor spacing, pattern orientation, etc) were programmed into the inertial navigation and Loran sets and the WRCS by the weapon systems officer. The Igloo White mission required extensive planning, as well as much extra pre-flight time in the 120°F temperatures common inside the aircraft revetments on the Ubon flight line.

The Loran-D added yet another external antenna and static discharge elements on the wing and tail trailing edges to the already cluttered shape of the F-4.

The WSO's view, never good, was practically eliminated by the addition of the Loran system. A KB-18 fore-aft, horizon-to-horizon panoramic strike camera was installed in the right forward AIM-7 Sparrow missile well. This was used to locate precisely the sensor release point and, through subsequent ballistic analysis, the impact point. The resulting precise location of the sensor was crucial to the correct interpretation of its data.

The acoustic sensors were dropped from the large SUU-42 dispenser, which was carried on one of the F-4's outboard wing pylons. The seismic sensors, with their peculiar "tree branch" antennae, were simply carried on the standard multiple or triple ejector racks (MERs/TERs).

Above: Posed photo with the second production RF-4C, 63-7741. Crew is Jack Krings, on ladder, and B. A. McIntyre. Somehow the angular new nose, with scoops and viewports which would have disfigured another aircraft, looked just right on the Phantom. *(MDC)*

**Above:** RF-4C in modern colours at Bergstrom AFB, Texas, during 1980. External tanks hold 370gal each. Bars on nose, fuselage, side and vertical stabiliser aid in formation flight. One of the hazards of night flying in Vietnam was for many years the lack of such aids; vertigo was common and sometimes disastrous. *(Jay Miller)*

**Below:** The view of the RF-4C most often seen by hostile forces, and then only briefly. *(Jay Miller)*

The electronic fence thus established across the lines of communication was attuned to a variety of signals, ranging from the noise and vibration of a truck moving to the jangling of personal equipment or the sibilant whisper of North Vietnamese conversations. This huge effort – which probably cost close to $8 billion in all – was intended to stop the movement of backpack-laden individuals and 2½-ton trucks, traffic which could have been throttled at its point of origin for far less cost, far less risk and far less damage to a nominally friendly countryside.

#### Close air support

Books have been written about air combat in every war: the account of von Richthofen's victories and defeat has been told and retold, as have the stories of all the aces of the Second World War, Korea and Vietnam. But there is another, even more dangerous task, that of close air support, and rarely have these crews received their just deserts. Close

air support was probably more important in Vietnam than in any previous war because the Vietcong and North Vietnamese usually held the initiative as to when to attack, and because there was no recognised "front line" or "bomb line" behind which friendly forces were safe. The enemy was everywhere, and the only way to protect small outposts or villages was by the application of close air support. In Southeast Asia the close air support units were on airborne or ground alert for calls to defend troops in battle (troops in contact, TIC) or to support a landing zone (LZ) extraction. Alternatively, they could carry out pre-planned strikes against "soft" targets, those with no elaborate ground-to-air defences.

Most of the missions were flown during daylight hours, though occasionally night strikes would be called. There were dozens of different ordnance mixes, each tailored to a particular mission. The following are typical:

With one 600gal centreline tank:
- Six Mk 82 500lb high-drag bombs and either two BLU-1/BLU-27 Napalm B bombs with fins or four without
- Six Mk 82s with two CBU-2/CBU-12 cluster bomb units; the latter were always mounted on the outboard wing stations
- Four BLU-1/BLU-27s without fins and two CBU-2/CBU-12s

With two 370gal wing tanks instead of the centreline tank:
- Four or six Mk 82s, and one SUU-16/SUU-23 gun pod on the centreline
- Four or six Mk 82s, and either two BLU-1/BLU-27 with fins or four without
- Four BLU-1/BLU-27s without fins, and one SUU-16/SUU-23 gun pod

Similar weapons would be carried at night, with the exception of alert duty, when two SUU-24 flame dispensers would be carried on the aircraft centreline.

When carried out by alert units close air support missions were normally handled by flights of two aircraft; when missions were pre-planned the flights occasionally grew to three or four. If the target called for more than two sorties additional flights were launched, usually at 20min intervals.

Some of the toughest missions were LZ extractions, which were flown in all areas, including North and South Vietnam, Laos and Cambodia. The task was to suppress ground fire while helicopters flew in and out of the contested area. The number of helicopters varied from one or two if a road-

**SA-2 en route to the target.**

Missile explodes beneath the aircraft, which bursts into flames.

Fuel flames and streams, aircraft begins to tumble, veer and break up.

Blazing wing breaks away.

Burning nose section tumbles earthwards.

This series of photographs depicts the fate which all too often befell Phantoms in the hostile skies of North Vietnam. On August 12, 1967, a 432nd TRW RF-4C flying from Udorn Royal Thai Air Force Base was destroyed by an SA-2 missile. Pilot was Lt-Col Edwin L. Atterbury, weapon system operator Maj Thomas V. Parrot. Both men survived the incident but Colonel Atterbury was murdered by his captors. *(USAF)*

watch team was being taken out, to dozens when it was necessary to move a large ground unit operating close to a superior hostile force.

The F-4s would lay down CBU-12s to generate a smokescreen before the helicopters landed, added to the screen just before take-off, and generally shielded the operation from observation or fire. The remaining ordnance would be used according to the tactical situation, under the direction of FACs or of the commander of the extraction, who would be aboard the choppers.

The most enjoyable and rewarding missions for most F-4 pilots were those flown from CAS (close air support) alert. Each fighter wing stationed in South Vietnam kept a portion of its aircraft alert to respond to immediate Army requests. Although this number was usually six aircraft, it often increased to 12 or more during periods of heavy ground activity such as the Tet offensive of 1968 or the 1972 spring offensive. Crews were on alert for 12hr periods or until they had flown three sorties, whichever came first. It was not unusual to fly three sorties while on day alert, but crews rarely flew more than two at night.

On going to alert status, the crews would pre-flight their aircraft, receive a current intelligence briefing, and then stand by in the squadron or special alert facility for instructions to launch. The latter normally gave rendezvous details, aircraft callsigns, radio frequencies and the type of target. A large proportion of the missions (perhaps as much as 90 per cent) were flown against "Vietcong in the open". These missions required extra care because friendly troops were usually close to the guerrillas. It was easier to support mechanised infantry or US Special Forces camps because then the friendlies were well protected in armoured personnel carriers, tanks or bunkers, allowing greater freedom in the use of ordnance. In any event, CAS missions usually resulted in tangible results, and consequently were highly satisfying.

**Armed reconnaissance**

Armed reconnaissance flights were designed to strike targets of opportunity along known supply routes: rivers, roads and rail lines. Missions were flown against North Vietnam, Laos and Cambodia, mostly in daylight but sometimes at night. The mission would be carried out by flights of two or four aircraft, depending upon requirements. Flights of two were easier to control, but more aircraft meant more eyes to search for targets and more ordnance to hit them with.

There were no typical weapon loads. The Phantom units were able to match weapons to targets, thanks to American engineering and logistics and the hard work of the tough

**An F-4 from Udorn racing in to Hanoi. Note the white streak of a Shrike missile trail below.** *(Fred Olmsted)*

young crews who toiled anonymously in the munitions areas. There were constant pressures to change loads: new commanders had theories to try out, and new threats and enemy tactics were always evolving. There was as much art as science to the technique, and this remains so to this day.

Night missions might call for two 370gal wing tanks and four LAU-3 rocket launchers (19 2.75in rockets with high-explosive heads), or six Mk 82s and one CBU-2, or two SUU-24 dispensers, each with eight flares. Day missions might have the same fuel tanks but a variety of stores that could include 750lb M117 general-purpose bombs, CBU-24/29 cluster bombs, or a 20mm gun pod.

The aircraft would launch to a designated area, with the lead carrying out a visual reconnaissance of selected routes. The lead would then indicate targets to other flight members, either by expending his own ordnance or by talking them in by reference to landmarks, just as an FAC would do. Flight members would attack under the direction of the lead, who could either put everything on one target or go on to other targets if there were enough tankers and munitions available.

Night missions were usually flown to a specific point on a supply route, where flares would be dropped and ordnance expended. The target area was picked out either by aircraft radar or, later, with the inertial guidance systems.

Armed recce did not give the same return on investment as other missions because visual reconnaissance by the flight leader was difficult and the ordnance loads cut down on aircraft endurance and manoeuvrability. When Fast FAC came into vogue armed reconnaissance was seldom flown.

### MiGCAP and strike force

The premier mission was of course MiGCAP, the positioning of F-4 flights between potential MiG attacks and the strike force. Curiously, the USAF had to relearn a lesson from the Second World War to make the strike forces truly effective.

The standard 16-aircraft F-105 bomber force was initially protected by two elements of four F-4s, one preceding the formation and one following it, and both weaving back and forth. The weave cost fuel, and as a result the F-4 pilots were instructed to stay close to the strike force and simply turn the MiGs away rather than engaging them. The Luftwaffe made this same mistake during the Battle of Britain. When the MiGs adopted low-level, rear-quarter hit-and-run tactics, however, the F-4s were given freedom to engage the North Vietnamese fighters on their own terms, and this proved to be far more effective.

The strike force eventually grew to be an awesome unit comprising many types of aircraft. The core remained bombers, with as many as 32 F-4Es carrying a mixture of conventional and smart bombs. "Iron Hand" flights, either F-105G or F-4C Wild Weasels and F-4Es carrying Shrike missiles, were used to suppress SAM sites. Vought A-7 chaff droppers preceded the formation and theoretically blanked out the SAM radars. Some felt that the chaff bombers actually put a signature on the formation, however. F-4s would range widely as MiGCAP, hoping to find a target. Finally, RF-4Cs would follow up the strike to assess the damage.

In addition to the actual strike force there were a host of ancillary aircraft without which it could not have functioned. The KC-135 tankers were fundamental to the operation and provided a safety margin after the strike.

Airborne command and control aircraft (EC-121D College Eye during the pre-1972 period, EC-121T Disco after) kept track of both friendly and unfriendly targets. Douglas EB-66 Skywarriors provided electronic countermeasures support to supplement the pods carried by the fighters later in the war. The Navy offered support with its own command and control and ECM aircraft whenever possible.

An immeasurable morale-booster was the search and rescue support provided by the force of Lockheed HC-130Ps, for control and refuelling, Sikorsky HH-53 Jolly Green Giants, and Douglas A-1 Skyraiders, which beat up the area around the downed crewman. This sophisticated and highly effective service had begun with three officers and three airmen in Saigon in 1962; by the time the war ended it had grown into the most competent rescue force ever seen and had saved 3,883 lives. During that period 45 rescue aircraft were shot down and 71 airmen killed, a bitter but unavoidable price to pay.

The strike force was the epitome of the American military effort to overcome its political handicaps with skill and science. In December 1972, in conjunction with the Navy and the B-52s, the strike forces proved all the theories to be wrong: after ten years of frustration the air was allowed to be the decisive weapon, and it was.

The high-g glory of air combat, the terror and elation of close air support, the satisfactions of a successful reconnaissance: these are one aspect of air combat. But the airmen's war had another side: the boredom of sitting in an alert shack; the discomfort of temperatures and humidity levels beyond previous experience; inadequate diet; the inevitable tendency to drink too much of liquor that was too good and too cheap ($2.00 for a quart of good whisky or brandy); the perils to home life; the sense of futility at fighting an unseen enemy who seemed to thrive on bombs; the growing awareness of civil resentment at home; and the certainty that American politicians had written off the war as unprofitable.

These irritations were compounded by certain real shortcomings in the early years of the war. There was misallocation of aircraft and pilots: squadrons would sometimes have 30 pilots to fly six aircraft. The GIBs were fighter pilots condemned to beg for rides like some "base-weenie," never getting to make a landing, never getting to pull the trigger, but always required to be ready, sweating in 20lb of flying suit, survival gear and parachute harness.

There were maintenance problems, too. The insulating compound in the electrical systems of the F-4 reacted in an unforeseen way to the temperature and humidity, breaking down into its component substances, leaving connectors loose, and flowing into areas where it caused further troubles. Bombs were often in such short supply that missions were flown with most racks empty (sortie rates were maintained for statistical purposes, in competition with the Navy). Food was bad, and Vietnam and the Vietnamese, after three decades of war, were cheerless and hostile.

The following extracts from the diary of Lt (now Maj, USAFR) Ronald W. Gibbs, a GIB with the 558th TFS, the Hammers, at Cam Ranh Bay from March 10, 1969, to March 10, 1970, convey the flavour of life for the average Phantom crewman in Vietnam:

**April 22, 1969** 9th mission. I Corps, 15 degrees 23 minutes north, 107 degrees 39 minutes east; mainly a road-cut mission.

**April 29** Couldn't believe the number of roads in the area that the VC use to get into Vietnam. Got two road cuts and a bunker. Main song around here goes: "We gotta get out of this place, if it's the last thing we ever do; we gotta get out of this place and make a better life for me and you."

**April 23** Got CARE package from Mom and Dad. Spent most of the day working on Hammer Inn [the squadron lounge]. Heard that Friday a four-ship is supposed to go to Steel Tiger in Laos and drop bombs. Bad news – won't have any guns, be just like the recon birds. Phu Cat lost 3 birds yesterday, 1 F-4, 2 F-100s. F-4 just exploded. All 3 in Laos. Bad place to go, worse than South Vietnam. Wish this whole thing was over. Going too far north to suit me.

**Below:** The oft-modified nose of the F-4 took on yet another shape with the installation of the M161A 20mm Gatling gun, capable of firing 6,000 rounds per minute. This crash programme proceeded almost flawlessly from start to finish. Shown here is the No 11 production F-4E. *(MDC)*

What's the matter with Da Nang? Why can't they go? Sure are closer than us, and we still don't have planes, only six flyable. Can't see why we can't get planes from the States – Davis Monthan, George and MacDill still have plenty.

**May 14** Alert all day. 16th mission. 50-caliber site; got 5 bunkers destroyed. Got 2 letters from Bitsy explaining situation at home. Sent her a four-page letter. Went to a movie but it was no good. Just watched TV.

**May 16** Alert and two more missions, last one was 20th, Air Medal No 4. 750s and CBUs. 1st was AAA site; got 10 bunkers, 1 AA site and gun destroyed, 1 AA site and gun damaged. 2nd was troops in contact with VC in woods. Got some KBA 'cause Army FAC said "Arms and legs all over".

**May 22** Flew off of night alert 0700; scrambled to a LZ prep near Cambodia. Slept most of the day and played pinochle at night. Got to get tape out tonight. Got drunk – Maitai; thinking about going to Australia on R&R or Hong Kong with Caroline in August.

**September 21** Alert. 88th and 89th missions. First was 25 miles west of 71 ravine; up a draw; had napes. Second just a show for Army, but did get some BDA. . . . Get front-seat ride tomorrow, I hope.

**September 22** Was supposed to have my first front-seat ride today, but due to lack of aircraft I was cancelled.

**February 5, 1970** Great day, had my last flight. 141st mission, with Jim Chapman. Target was usual – nothing. We did victory rolls over the field in burner, what a sight it must have been. I rode the aircraft back in with a WW II flying act. Felt like a big weight gone!

Gibbs' experience was undoubtedly more typical than that of aces like Steve Ritchie or Chuck DeBellevue. His 12th TFW had been one of the first F-4C units to arrive in South Vietnam, and consisted of the 557th TFS (Sharkbaits), 558th (Hammers) and 559th (Billygoats). By the time Gibbs arrived, many of their aircraft had been inherited from Robin Olds' 8th TFW and had been in South-east Asia since 1965; some still bore red stars from the victories in the North. The potting compound problem made it not unusual for a pilot to try to make a radio transmission only to have his speed brakes deploy instead. Crews sometimes found that after selecting a ripple drop they had jettisoned everything, including drop tanks and bomb racks. It was a strange situation, and the aircraft were finally flown singly, limited to 1½g maximum and 30° of bank, to Japan and the Philippines for repair.

Like almost all the pilots, Gibbs was eager when he arrived. When F-4 missions became hard to come by he volunteered to become an OV-10 FAC, but the wing commander vetoed the request. In the twelve months he flew with the 558th the unit lost only four aircraft and recovered seven of the eight crew members, a remarkable record. One last entry from his diary, this time from October 22: "Dick Hickenbottom and I on a road mining mission in Laos. 'Gomers' dug in with 37mm cannon – they'd roll out, snap a shot, and roll back in. Frag prescribed 425kt attack speed, but went in at Mach 1 – flew under tracers and got out." In December, another 558th crew, Lts Ben Danielson and Woody Bergeron, made an attack on the same target and were promptly shot down. Both landed right amongst the

**Left: A St Louis test bird with EROS anti-collision package and Sparrow and Falcon missiles.** *(MDC)*

**Below: Introduction of wing slats under the "Agile Eagle" programme led to a reduction in the flap angle used for final approach, and also altered the airflow around the tailplane. The solution was to call for only 30° of flap and a slight increase in approach speed (about 8mph), and put slots on the tailplane leading edges.** *(MDC)*

enemy, one on each side of a river. Search and rescue tried for seven hours to pick them up but were driven off by heavy enemy fire. The next day the enemy caught Danielson, who resisted with a hand gun, like Frank Luke of the First World War, until he was killed. Bergeron spent three more days on the ground and then was picked up. He was so certain that he would be rescued that he had carefully saved samples of the ground water he had been drinking for later analysis to determine if he would need any post-rescue medical treatment.

Gibbs wanted very much to continue in the Air Force and become a front-seater in the F-4. The Air Force did not freely provide that option, however: the crewman had to volunteer for another tour in South-east Asia or else be assigned as a co-pilot elsewhere, on strategic bombers or transports. Gibbs finished his tour of service and got out, and is now a successful banker in Blacksburg, Virginia.

The war that Ronald Gibbs fought must be familiar to most USAF Phantom aircrew of the time, but occasionally there were missions that brought enormous satisfaction to the participants. Lt Harry Brown was a back-seater with Satan's Angels, the 433rd TFS of the 8th TFW, whose calling cards modestly proclaimed them to be "The World's Greatest Fighter Pilots," "Modest Heroes," "International Lovers" and "Masters of the Calculated Risk". (Other squadrons had similar cards, with such embellishments as "Alligator Wrestlers" or "Fastest Gun in the South-east".) Brown was stationed at Ubon from October 1971 to October 1972, and his aircraft had been equipped with Loran and Paveway and Pave Knife laser designation systems. He flew over 40 Loran missions, two of which stand out in his memory.

The first of these was against the Long Chi power station in April 1972. This new hydroelectric facility was a tough target because it was located within 100ft of its associated dam. Dams were off-limits to US air power because of the damage to the enemy countryside which might result if they were breached. Thus they became a good place to store valuable equipment or to locate SAM or AAA sites.

Col (later Brig-Gen) Carl Miller was wing commander of the 8th at the time, and he led three four-ship elements of F-4Ds in the strike force. The usual complement of MiGCAP and Iron Hand flights accompanied them. The first and last four-ship elements, from the 433rd TFS, were targeted on the power station itself, while the second element, from the 25th TFS, was to hit the transformer yard.

**Above: Well armed with a mixture of fuze-extended bombs, an F-4E of the 421st TFS, 366th TFW, taxies out at Da Nang.** (Lt-Col Oleg Komanitcky)

**Top: "Agile Eagle" test vehicle. A beryllium rudder was also tested during Agile Eagle.** (MDC)

Miller was leading, with Lt Wayne King as his GIB, and Brown was in the back seat of Lt-Col Rick Hilton's F-4, the lead ship of the third element.

Each of the element leads had Pave Knife equipment for designation of targets for the laser-guided Mk 84 bombs. Pave Knife acquired the target with a laser beam which could be controlled for target-seeking purposes from the back seat. The GIB had a 5in Sony television screen which enabled him to direct the beam on to the target. The beam then provided an electronic aiming point, with a notional funnel or "basket" down which the laser-guided bombs (LGBs) would be directed. (Targets could also be designated by other aircraft – the AC-130, for example – though not in high-risk areas.)

Miller rolled in on the target, and King, using the little TV set, directed a 2,000lb Mk 84 (worth about $5,000) directly onto the power station. The blast blew the roof off, for the fuzes had been set for a 0.025sec delay to ensure at least 20ft penetration before detonation. The second flight, to its eternal embarrassment, missed the target completely, but Hilton and Brown's four-ship came in to put their bombs inside the now roofless plant, utterly destroying it.

The 23mm and 37mm flak had been intense, but the Pave Knife technique of entering a dive, releasing a bomb at about 14,000ft and pulling out at 11,000ft had kept the Phantoms above most of the gunfire while still vastly improving accuracy.

Brown's second favourite mission was against that *bête noire* of US air power in Vietnam, a bridge. The traffic essential to operations in the South flowed over bridges, and they were always protected by the heaviest flak and SAM defences. When the bridges were hit, the resourceful North Vietnamese would rapidly rebuild them, sometimes submerging the rebuilt areas so that it appeared that no work had been done, at other times investing the time, steel and concrete necessary to make them almost impervious to ordinary bomb hits.

More than 1,000 sorties had been carried out over the years in an effort to destroy the Thanh Hoa bridge, which carried the main-line traffic south from Hanoi; over 30 F-105s had been lost in these attacks. But still it stood, a massive concrete and steel structure heavily protected by AAA and SAMs. It had been damaged in April 1972 by F-4s with Mk 84 LGBs, and on May 13 a flight of two F-4s finished it off. Lt Jack Smith (who went on to become a leader of the USAF's Thunderbirds demonstration team, and who tragically died in an accident in Cleveland) was in the front seat of the lead aircraft, with Brown in the back seat. The No 2 aircraft was flown by 1st Lt John Creech, with back-seater 1st Lt John Cole, who would be killed in action later in the war.

The weather was bad, about 5/8ths overcast, but then the clouds opened up a little to reveal the bridge. Brown acquired the bridge on his Sony, and the two-ship team smothered the span with 2,000lb Mk 84s and a single 3,000lb "Fat Albert" Mk 118. Two aircraft with LGBs had done what a thousand could not.

Between the glamorous and the boring, between the tedious and the satisfying, there was always the common denominator of danger and, all too often, death. Reproduced below are the last radio transmissions of the final flight of RF-4C Strobe 10, which took place on July 23, 1968. The aircraft commander was Maj-Gen Robert F. Worley, vice-commander of the Seventh Air Force, Saigon. In the back seat, as instructor pilot, was Maj Robert Brodman. The language is sometimes salty, as it often is during air combat. Strobe 10 had been hit by ground fire north-west of Da Nang.

**Maj-Gen Robert F. Worley, who was killed in the crash of an RF-4C shortly before he was scheduled to leave Vietnam.** *(USAF)*

| Transmitter | Content |
|---|---|
| Unknown | Be advised that Strobe 10 is working on Guard [listening frequency] with Crown 4 [airborne command and control post]. |
| Unknown | Roger. |
| Misty 31 | Strobe 10, Misty 31 [Fast FAC aircraft] on Guard. |
| Strobe 10 | Go ahead Misty, this is Strobe 10. |
| Misty 31 | OK, I am coming up on your left wing, I believe. You want to check me? |
| Strobe 10 | Roger, how about giving us a check. I'm getting a lot of smoke in the rear cockpit. We've dumped but it's still coming in. We still have hydraulic pressure, however. |
| Misty 31 | OK, is this me off your left wing? |
| Strobe 10 | That's affirmative. I wish you'd take a look at our underside. I'd really appreciate it. That's where we took the hit. |
| Misty 31 | OK, stand by. |
| Crown 4 | [To rescue helicopter] OK, maintain your present position. I'm going to run Strobe down past you. Strobe 10 is at Flight Level 210. |
| Rescue | Thankyou. |
| Waterboy | Strobe, this is Waterboy. [Waterboy was a control and reporting post near Dong Ha] |
| Strobe 10 | Goddamn, what's wrong with this . . . airplane. [(Weakly, apparently from other seat:) I don't know.] Say again, can we turn right now to 130? |
| Crown 4 | Negative. Turn right to 115, down the coast. |
| Strobe 10 | Roger, 115. |
| Crown 4 | Waterboy, how do you read? |
| Misty 31 | Strobe, what's your airspeed? |

| | |
|---|---|
| Strobe 10 | Roger, right now it's 300 knots, and I'm getting a lot of smoke in the rear cockpit. |
| Crown 4 | Strobe, I recommend you get down below 200. You'll be in the soup if you have to punch out. |
| Strobe 10 | OK, we got 11,000 now. |
| Crown 4 | OK, copy. |
| Waterboy | OK Crown, I'm heading Strobe right directly for you. He's down at 11,000. |
| Crown 4 | Roger, we copy. |
| Waterboy | He's presently 12 miles north-west of you, heading south. |
| Misty 31 | OK Strobe, Misty overlooking you here. |
| Waterboy | Jolly 28, start heading down the coast. |
| Strobe 10 | Yeah, OK. This . . . slowed down. |
| Strobe 10 | [Apparently from back seat to front seat] Hey, we got a fire. You got a fire up there? |
| Misty 31 | Hey Strobe, you got a fire underneath the rear cockpit. |
| Strobe 10 | What is it now? |
| Misty 31 | Strobe, straighten your wings baby. You got a fire there underneath the back cockpit. |
| Misty 31 | You got a fire in your nose section and all the way back to the rear cockpit. It's clear up in your camera bay. |
| Strobe 10 | OK, we are going to have to bail out. We are going to get down and go ahead and bail out. |
| Waterboy | OK Strobe, we got contact on you. You're clear to bail out when you have to. [Much heavy breathing] |
| Waterboy | Strobe 10, Waterboy. |
| Strobe 10 | Strobe 10 bailing out now. |
| Waterboy | OK, we got contact, marker beacon. |
| Crown 4 | 10, Crown 4. [Sound of emergency beeper] |
| Misty 31 | Beautiful, the back-seater's out. He's got a good chute. |
| Crown 4 | Waterboy, vector to 097 at 24. |
| Misty 31 | And the front-seater hasn't bailed out yet. Get out, goddamn. |
| Misty 31 | Strobe 10, bail out, bail out. |
| Misty 31 | Strobe, this is Misty, bail out, bail out. |
| Misty 31 | Strobe 10, bail out, bail out. |
| Crown 4 | Waterboy, stay with the back-seater. |
| Misty 31 | OK Crown, Strobe is heading inbound. Airplane just blew up. |
| Jolly Green 28 | Jolly Green off your six o'clock. |
| Misty 31 | Strobe 10, bail out, bail out. Come on you son of a bitch. Oh god almighty no, look at him burn. OK, cockpit's on fire and the airplane is going in, canopy still on it. [Heavy breathing] |
| Crown 4 | OK, what's your position, Misty? |
| Misty 31 | Just gone feet dry; going to hit the ground any second. |
| Misty 31 | [Intercom comment] Keep the altitude up, goddammit. |
| Misty 31 | Oh my god, the airplane just impacted on the beach. |
| Crown 4 | OK, say your position. |
| Misty 31 | Squawking mayday. |
| Misty 31 | [Intercom] Watch the helicopter, 12 o'clock level. |
| Waterboy | Jolly 28, do you have the No 2 man? |
| Misty 31 | Son of a bitch, isn't that terrible? |
| Crown 4 | Misty, how about going down to take a look? You think there is any chance for survival? |
| Misty 31 | Negative survival, negative survival. |
| Crown 4 | Crown copies, OK. |
| Misty 31 | [Intercom] He was dead before he hit. |
| Waterboy | Jolly 28, Waterboy. Do you have the back-seater in sight? |
| Misty 31 | I'm orbiting the back-seater's position now. The canopy never came off the aircraft, and there was the cockpit full of fire and smoke. [Intercom (aircraft commander): Did you see that fire up there in that forward electronics bay? It was right under the back-seater. (GIB) Negative, it was all the way up in the camera compartment. The time was 12.05 wasn't it? (AC) Yeah, the back-seater hit the water about 12.10. Hey, the Jolly's right there isn't he? That guy probably won't even be out of his harness yet.] |
| Misty 31 | Hey Crown, Misty 31. |
| Crown 4 | Go ahead Misty. |
| Misty 31 | You got a junk going toward the survivor, I'm going down and take a look at him. |
| Crown 4 | Rog. What have you got on board? |
| Misty 31 | I can fix him if he's hostile. I don't want him over there anyway. I'm going to come across his bow. |
| Crown 4 | Roger. What do you have for ordnance? |
| Misty 31 | 20mm. I'll have him if he's bad. |
| Misty 31 | [Intercom: You got the guns armed? Yeah.] |
| Misty 31 | [Intercom (aircraft commander): They're not friendly at all. We are going back and put some 20mm in front of them. You got the airplane? Fire some 20mm right across the bow. (GIB): OK, I got the airplane. I think they got the message.] |
| Crown 4 | Crown copies. |
| Misty 31 | [Intercom (aircraft commander): He turned around, didn't he? (GIB) Yeah. Keep an eye on the son of a bitch. He's really moving.] |
| Misty 31 | [Intercom (aircraft commander): Wonder if the rear seat going had anything to do with those flames picking up. (GIB): Yeah, it could have, you know.] |
| Misty 31 | [Intercom (aircraft commander): Yeah, it snuffed him right out. I'll bet that was it.] |

Maj-Gen Worley was the second Air Force general to be killed in combat: Maj-Gen William J. Crumm had been killed on July 6, 1967, in a mid-air collision of two B-52s. Maj Brodman suffered only minor injuries in the RF-4C

Above: **F-4D 66-8793 of the 25th TFS, 8th TFW, based at Ubon RTAFB, Thailand.** *(Capt Albert Piccirillo)*

Left: **This F-4D, 65-0637 Cobra 23 of the 480th TFS, 12th TFW, was shot down by ground fire over central Laos on February 25, 1971. The crew – pilot Capt Hedditch, WSO Lt McLaughlin – were picked up on February 26.** *(Norman E. Taylor)*

Right: **The Phantom's prowess in combat owed much to the USAF training effort. The first Air Force Phantom to log 1,000 hours was F-4C 63-7418, assigned to the 4453rd Combat Crew Training Squadron at MacDill AFB, Florida. Shown here are the men who flew the 1,000th hour: instructor pilot Capt George O. Watts (rear seat) and 1st Lt Joseph F. Schucter.**

ejection. As a result of Worley's tragic death, two important changes were made. First, general officers were no longer allowed to fly combat. Second, and perhaps more significant, F-4 ejection procedures were altered so that the aircraft commander would initiate ejection for both crewmen, allowing for the possibility of one being incapacitated.

The loss of Maj-Gen Worley was the first of three incidents which led to the pinpointing of a previously unknown problem with F-4 ejections. Capt Gene Tucker saw the reports on the second and third incidents. He was attached to VF-33, flying off USS *America*, when Lt J. G. Eric Brice's Phantom was hit by flak just aft of the nosewheel, in an area filled with pneumatic and hydraulic lines. The crippled Phantom managed to go "feet wet" and the RIO, Lt JG Bill Simmond, ejected, being severely injured in the process. The F-4J then did a lazy high-speed barrel roll into the sea, and Brice did not eject. This was a mystery, because he was fine before Simmond left. A few days later the squadron executive officer, Cdr Orville G. "Tex" Elliot, was hit on a mission. The aircraft was nursed to the sea and the crew set themselves up for ejection. The RIO ejected safely, and several seconds elapsed before Elliot pulled his face curtain. Nothing happened. He then pulled the canopy jettison handle, followed by the manual unlock handle. Still the ejection sequence failed to start, even though the face curtain had been pulled sufficiently. The aircraft was out of control and headed for the sea, but Elliot was a big, burly, barrel-chested man and refused to give up. He pushed on the inside of the canopy, meanwhile broadcasting over the radio that the canopy wouldn't come off, so as to give some indication of what had happened if he didn't get out. He heaved one more time and the canopy edged into the slipstream and immediately separated. Elliot then ejected safely.

Analysis of the three incidents indicated that with the rear canopy gone at high airspeeds there was too much negative pressure on the front canopy for the existing system to open and jettison it. Bow thrusters, small gas-operated pistons, were installed in the rails at the base of the front canopy to push it up into the slipstream and ensure that it detached.

Six months after the Worley incident, on January 25, 1969, the US and North Vietnam began a long and tortuous peace negotiation. The US wanted disengagement with face-saving, and in the hope that a nominal South Vietnamese state would be maintained. The North Vietnamese wanted the US out, followed by the establishment of one Vietnamese nation with the capital in Hanoi.

The US efforts were punctuated with periodic troop withdrawals, and accompanied by a process of "Vietnamisation". US troop strength in South Vietnam declined from 536,100 on December 31, 1968, to 474,000 a year later, 335,800 at the end of 1970, approximately 190,000 by the end of 1971, and 69,000 by May 1, 1972. Then, in December 1972, when troop strength was reported to be down to 27,000, the US defeated North Vietnam militarily by the use of strategic air power.

During the American withdrawal the North Vietnamese had systematically increased their strength in the South: they began the invasion of Quang Tri and Thuy Thien provinces, increased their "in-country activity" and stepped up incursions from Laos. Finally, in March 1972, they began a large-scale offensive using 40,000 regular troops and large numbers of tanks, heavily supported by flak and anti-aircraft. There were no longer enough ground forces in the theatre, and the US was forced to use air power as it had always been meant to be employed. Tactical air power was available almost solely from Thailand, while the B-52s could be employed both from Thailand and Guam.

American air power provided the stiffening needed to permit the Vietnamised South Vietnamese forces to hold on,

and prevented the sort of rout that would subsequently occur in 1975. Once again American technology – in the form of Loran, smart bombs and electronic countermeasures – provided the necessary edge. The Phantoms had a field day in air combat, shooting down 49 MiGs in 1972, with one final victory in January 1973.

One of the USAF's brightest young officers, Maj Robert A. Lodge, scored the first victory of this new period of the war when, on February 21, 1972, he and his equally famous back-seater, then 1st Lt Roger C. Locher, were flying MiGCAP in an F-4D. They were members of the 555th "Triple Nickel" TFS, now a part of the 432nd Tactical Reconnaissance Wing.

Red Crown, the famous Navy radar ship, had called out the location of a MiG group and Locher quickly locked on. The target was closing at a combined velocity in excess of 900kt and Lodge let loose three AIM-7E Sparrow missiles, which had the invaluable quality of being able to make a head-on attack. The second missile hit the MiG, and there followed the "customary fireball with its standard debris, hair and teeth."

On April 16 three more MiGs were brought down by 432nd aircraft. The first victory, Captain Fred Olmsted's, was described in Chapter 1. The second was scored by the crew of Basco 03 of the 13th TFS. Aircraft commander Maj Edward D. Cherry claimed his first victory, and so did back-seater Capt Jeffrey S. Feinstein, who would go on to score four more and become the third USAF ace in Vietnam. Cherry and Feinstein fought a gruelling engagement, following a MiG-21 through a whole catalogue of manoeuvres, firing off an AIM-9 without success, and then not being able to get in position to loose off any further missiles. Their wingman, Basco 04, steamed in and let fly four Sparrows, and the still lucky MiG managed to avoid them all. Finally, Feinstein secured yet another radar lock on the MiG while they were following it in an 80°-bank descending turn. The Sparrow blew the MiG's right wing off; the pilot ejected and Cherry and Feinstein flew within 500ft of the defeated pilot's parachute.

An F-4D from the 523rd TFS knocked down the third MiG-21 that day. Capts James C. Null (AC) and Michael D Vahue (WSO) in Papa 03 acquired an early radar lock on a flight of MiGs, but Papa 04 was cleared to attack. Unfortunately, all of 04's Sparrows failed and the silvery communist fighters passed overhead. Null pulled into a tight right-hand turn and positioned his F-4 for another head-on attack. This time Null fired three Sparrows, the third of which blew off the MiG's tail section.

The remainder of 1972 saw more and more responsibility placed on American aircrews. As the South Vietnamese ground forces were left increasingly on their own, and the North Vietnamese became bolder, the pressure for American air power to equalise the situation became greater. The build-up in North Vietnamese defensive capability in South Vietnam continued, with SA-7s, SA-2s and heavy AAA being encountered in the northern provinces.

The US response to the North Vietnamese spring offensive had begun as Freedom Train but was enlarged to Linebacker on May 10, when President Nixon resorted to bombing above the 20th parallel as a means of halting the collapse in South Vietnam. This was so successful that on

Above: A fighter pilot's card. There were many variations on this formula, most of them more ribald.

October 23 he was able to halt bombing above the 20th parallel once again, the North Vietnamese having returned to the negotiating tables in Paris. During this period the F-4s had achieved a further 37 victories, though at a high cost as a result of improved NVAF application of high-speed hit-and-run tactics. During June and July the exchange ratio had fallen to one to one because the MiGs were airborne for only 12-14min in the course of their six o'clock attacks.

New tactics, in concert with the Navy's radar capability, solved the problem, and the exchange ratio for the period August-October 15, 1972, reached four to one. If there had been an Awacs system, with its far-seeing lookdown capability, a large part of the USAF and Navy casualties could have been avoided.

The North Vietnamese resorted to their diplomatically effective delaying tactics again, and the United States was left with no option but to execute the bombing plan that had been first suggested in 1965. Linebacker II was put into operation on December 18, and for 11 days the combined weight of USAF and US Navy aerial power hammered North Vietnam. The North Vietnamese, shaken by the force of the attack, returned once again to the negotiating tables and on January 23, 1973, the ceasefire came into existence.

During this period the main weight of the attack was applied by B-52s, though the F-4s continued in their flak-suppression and bombing roles. Only four air-to-air victories were recorded, primarily because the MiGs had been denied the use of their airfields.

The issue had been decided by air power, even though the enemy had been given seven years to train, build up defences and devise tactics. In 1965 it would have been much easier for everyone – North Vietnamese, South Vietnamese and, certainly, the Americans. The accountants who had devised the elaborate rules of engagement which ensured that the war could not be won from the air were proved wrong. But in seven years the political situation in both Vietnam and in the United States had so changed that the result was no longer in doubt to either side. The United States would leave South Vietnam, and shortly afterwards the North Vietnamese would at last take over. Air power's victory – in which the Phantom played such a major part – had years before been discounted at the bargaining table.

# 5. Aces and issues

In the twelve years of involvement, from Farm Gate to the peace settlement, only five aces emerged from the Vietnamese conflict. Three – Capt Chuck E. DeBellevue (six victories), Capt Steve Ritchie (five) and Capt Jeff Feinstein (five) – were from the USAF; two, Lt Randy Cunningham (five) and Lt Willie Driscoll (five), were from the US Navy. There are two major reasons why this total is so low: first, the MiGs were often hard to find and difficult to engage; second, the US pilots were rotated through their combat tours so that experience was spread wide rather than deep.

The statistics on these aces are significantly different from those applying to their predecessors in earlier wars. While it appears that they accounted for a total of 26 MiGs, the actual number shot down by these men was 17. Behind this oddity lies a decision by the late Gen John J. Ryan, Chief of Staff of the USAF, and his US Navy counterparts that both the pilot and the GIB would receive full credit for a victory, rather than sharing. This was done mainly for morale reasons but also from a sense of fairness. The Phantom was a complicated aircraft, demanding teamwork. And, as has been noted, the Vietnam War was not popular, and it seemed certain that there would be no aces to appeal to public sentiment if credit was shared fifty-fifty.

Understandably, there were two strong opinions on this matter. Pilots almost uniformly rejected the idea of a radar intercept officer, a weapon systems officer, a GIB or anybody else not in the front seat receiving a share of the victory. The GIBs, to a man, felt that their contribution was so great that they deserved equal credit. But the ace question was only one part of an even more heated debate: whether a fighter aircraft should even have two crewmen. This was settled in the acquisition process, but then another issue arose: should the second man be a pilot, a navigator, or what?

The Phantom II was designed with two seats because the electronic gear of the time was so complicated in operation that a second man was requird. As George Spangenburg noted in his report of November 28, 1958, "Under adverse

**Ace's eye view. The pipper is on the MiG's port wing, the F-4 pilot fires . . .**

**. . . and the enemy fighter's wing tank blows.**

**Right: Capt Steve Ritchie in the front seat and Capt Chuck De Bellevue in the back give the classic thumbs-up sign. Ritchie was the first USAF ace, while DeBellevue went on to record six victories.** *(USAF)*

**Centre right: The qualities that are essential in an ace show up here: confidence, pride, a touch of arrogance, good humour. Steve Ritchie after his fifth kill, which made him the first USAF ace.** *(USAF)*

**Far right: Capt (now Lt-Col) Charles "Chuck" DeBellevue, the highest-scoring American ace of the Vietnam War.** *(USAF)*

conditions of very high closing speeds, atmospheric conditions or ECM environments that result in short detection ranges, earlier detection by only a few sweeps of the radar can be the difference between success and failure."

As the weaponry of the F-4 became more complicated, and the conditions of combat more restrictive and complex, the man in the back seat became even more essential. Against this logic went the desire of many fighter pilots, even including pilots who flew in the back, not just for one seat but for only one engine. Fighter pilots are by nature independent doers who want always to be in control; they are risk-takers, but don't want to hazard others or be hazarded by them in turn. Studies have shown them to have certain characteristics: aggressiveness, confidence, self-discipline, dedication, awareness, an instinct to hunt and kill, courage, fast reactions and so on. The pilots pride themselves on possessing these qualities, and believe that when it comes to a fight they cannot be added to by another man in the back seat.

The Navy approached the problem with the firm idea that the man in the back seat, the radar intercept officer (RIO), was to supplement the pilot by providing him with an extra set of eyes and hands, and with specialised knowledge of how to get the most from the complicated equipment they shared. The Navy went a sensible step further by trying always to keep crews together. The pilot and his RIO were seen as a team who should fly together on every occasion, in the same aircraft when possible, and on the same types of mission. The pilot had a basically offensive role, the RIO a defensive one.

The USAF had a totally different approach, and when it selected the Phantom (the F-110 at the time) it decreed that the back-seat position should be modified to include a full set of controls so that two pilots could operate the type. There were some apparent safety and efficiency advantages to this idea, and it seemed at the time logical that the aircraft commander would be a senior pilot who would train a younger pilot, just out of flying school, in the intricacies of air combat. The young man would gain experience and eventually move up to the front seat.

Two factors worked against this approach. First, the front-seater often did not himself have enough flying time and experience to fly the Phantom to its limits, and he was unwilling to give up any of the experience – in-flight refuelling, combat, landings, instrument flight – which was to be gained in a mission. Secondly, the Air Force did not succeed in maintaining the crew integrity that was so necessary. Lt Ron Gibbs (see Chapter 4) complained rightly that he was never permanently assigned to an aircraft commander, which interfered with his training. He had to prove himself to each new man in the front seat, just as that man had to prove himself to Gibbs. This was repeated over and over again, with an inevitable loss of co-ordination and confidence. Part of the problem lay in the fact that many of the older pilots refused to acknowledge the new man or even to communicate with him.

Air Force GIBs suffered further indignities in South-east Asia. Promised some sort of a programme under which they could anticipate moving from the back seat to the front, they found themselves required to volunteer for an immediate return to a combat tour in exchange for aircraft commander training. To add insult to injury, the accelerating pace of the war resulted in freshly minted products of undergraduate pilot training appearing as aircraft commanders. GIBs with perhaps a hundred combat missions were assigned to a new pilot whom they would have to train. It was intolerable for many.

For many GIBs the last straw, the one that made them leave the Air Force, was what happened after a tour in F-4s in South-east Asia. Instead of going to a fighter squadron they would be assigned to Strategic Air Command or Military Airlift Command, where they would now find themselves GIRs ("Guys in the Right Seat") in the B-52, KC-135, C-141 or C-130. The Phantom GIB might have more total time than his new aircraft commander, who had probably never been in combat, yet he would lack the required 500 hours in the type to get into the left-hand seat.

The expanding war led to a greater demand for Air Force F-4 back-seaters than pilot training could provide, so it was decided to put navigators in the back seat. They were sent on a short transition course – ten sorties when a hundred would have been reasonable – and then sent into combat.

Steve Ritchie was the first USAF ace of the Vietnam War, shooting down five MiG-21s with missiles. He claimed four of the aircraft with Chuck DeBellevue as his back-seater.

**Above: The six-starred aircraft in which Steve Ritchie and Chuck De Bellevue got their combined kills. Ritchie stresses the need to keep crews together and consistently using the same aircraft wherever possible. The photograph was taken later in the career of the aircraft, at a base in Florida.** *(MDC)*

DeBellevue, who went on to score two more victories, is the leading ace of the war, with a total of six. Ritchie and DeBellevue were partners, comrades, friends. They admired each other's ability then, and respect each other's opinions now, even though they are sharply opposed on the issue of back-seat aces.

After becoming an ace Ritchie was debriefed by Gen William W. Momyer, commander of 7th Air Force from July 1, 1966, to July 31, 1968, and later commander of Tactical Air Command. Ritchie told Momyer that it was advantageous to have two men in the F-4, but under two unbreakable conditions: both crew members had to be highly qualified, and they had to fly together all of the time.

In Vietnam, Ritchie recalls, there were only a few aircraft commanders actually qualified to go into Route Package VIA*, the Hanoi area, and as a rule the back-seaters were even less qualified. The level of training received by the navigators was so low that many even had trouble interpreting their radar scopes. As a result, after combat missions the Phantom crews would tank up and go to an area in Thailand so that the pilots could receive basic combat formation training while the navigators practised on their equipment.

*For reasons of control, North Vietnam was divided into seven "route packages," I, II, III, IV, V, VIA and VIB. The USAF was responsible for I, V and VIA, the Navy for II, III, IV and VIB. Many people regard this geographical pie-slicing as strategically and tactically wrong.

The failure to keep crews together was one of the reasons for the declining F-4/MiG exchange ratio, according to Ritchie. An improperly trained back-seater, or one with whom the pilot was not familiar, was at times a hazard to the mission. Ritchie remembers a mission in which an inexperienced back-seater called out a flight of enemy aircraft. Ritchie immediately turned in reaction – into a flight of F-105s. This led his entire four-ship out of its strike force protection position and could have been disastrous if MiGs had been present in another quarter.

Not surprisingly, DeBellevue has significantly different thoughts on the matter. Chuck went on to pilot training after the Vietnam War and is now an F-4 squadron commander, firmly in the front seat. Here are his views: "The GIB was a new breed of cat not formerly known to the fighter community. It was the GIB, as an integral part of the F-4 crew, that helped to make the Phantom II the multi-role aircraft it was. Each different type of mission – air-to-air, air-to-surface, etc – involves separate and distinct tasks, and each mission places different requirements on the crew. The GIB, in particular, must be flexible enough to handle all these missions at short notice. [According to Steve Ritchie, Chuck DeBellevue was among the one in a hundred GIBs who could do this.]

"The GIB assists the aircraft commander (AC) in the accomplishment of any mission; he is a navigator, an electronic warfare officer (EWO), a radar intercept officer (RIO), a bombardier, a co-pilot; he is a jack of all trades who is as much responsible for the successful accomplishment of the mission as the AC.

"A great deal depends on the back-seater; his efforts can make the difference between a crew being average or really shining. If the GIB doesn't pull his load, it makes it extremely difficult for the AC to be effective.

"On a combat mission some of the important things for a GIB are: situational awareness; position; nearest safe area; direction to home; location of the threat; status of on-board systems; fuel status; assessment of damage, and so on. Coming out of an air-to-air engagement, little things – like the right direction to turn to get home, or a fuel check, or the tanker location – all become critical.

"The F-4, as a two-seat fighter, requires the efforts of both guys to be successful; there is no room for anyone who is just along for the ride."

The following accounts of combat from Ritchie and DeBellevue reveal what a high-speed dogfight looked and felt like to both pilot and back-seater.

Few men have been as well prepared for combat before scoring a victory as Steve Ritchie; fewer still had that combination of experience, training and the desire to get the maximum out of his aircraft that made him an ace. Ritchie first came to South-east Asia in 1968. On his first tour he flew 195 missions, 95 of them as a Stormy Fast FAC. He then returned for training at the Fighter Weapons School at Nellis AFB, Nevada, which was desperately trying to translate the lessons learned in Vietnam into tactical doctrine. He served as an instructor in the school and then returned to the war for another tour.

Assigned to the 555th TFS, 432nd TRW, at Ubon, he got his first MiG-21 on May 10, 1972, in a very complex engagement marred by the loss of one of the USAF's finest tacticians, Maj Bob Lodge. In a wild mêlée with four MiG-

**Above: Capt Jeffrey Feinstein, who became an ace while flying with the 13th TFS.** *(USAF)*

**Below: Looking more like a television hero than seems reasonable, Capt Fred Olmsted poses on the ladder of his aircraft.**

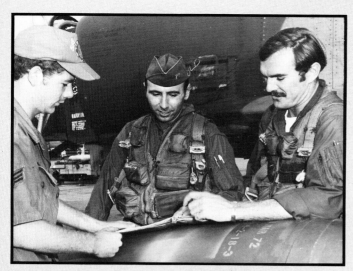
Olmsted (right) and WSO Capt Gerald R. Volloy after their AIM-7 kill of a MiG-21 on March 30, 1972. The crew chief, as so often is the case, is unidentified.

Olmsted and WSO Stuart W. Maas after Olmsted's second kill, scored on April 16, 1972, over Hanoi. *(USAF/SSgt Charlie R. Strutts)*

Left: Olmsted taking off with a load of CBU-52 cluster bombs. *(Fred Olmsted)*

Post-strike happiness: (left to right) Jeffrey Feinstein, Maj Edward D. Cherry, Olmsted and Maas celebrate Cherry and Feinstein's first kill, Olmsted's second.

Champagne and cigars were traditional at post-kill celebrations. *(Fred Olmsted)*

21s, Lodge and his wingman, 1st Lt John D. Markle, had knocked down two in a head-on attack, firing their Sparrows at long range after identifying the enemy with the classified Combat Tree equipment. The F-4s then turned on the remaining MiGs and Ritchie got his first from about 5,000ft with a Sparrow. DeBellevue was in the back seat. Lodge, fixated on the remaining MiG-21, ignored warnings about two MiG-19s which had attacked from six o'clock high and which hosed him with 30mm cannon fire. The man Ritchie calls, with DeBellevue, the "best GIB in SEA," Capt Roger Locher, bailed out but Lodge went in with the aircraft. (Locher, a veteran of 400 missions, probably the greatest number for any GIB in the war, survived for 23 days on the ground behind enemy lines before being picked up.)

Ritchie's second kill came on May 31 while flying F-4D Icebag 01. He let attacking MiGs race in from the 7-8 o'clock position, pulling up at the last minute to jettison the centreline fuel tank and then descending in a pirouette turn to a rear-quarter position as the MiGs overshot and split. He fired two Sparrows; the first corkscrewed off to the right as the enemy sliced to the left, so he fired two more. The second missile blew up half-way to the MiG; the third guided perfectly but blew up 500ft behind the fleeing enemy. The fourth went way out in front, correctly computing the required lead, and then slowed down as its engine burned out. The MiG pilot was probably peering back anxiously over his shoulder as the Sparrow slammed into his cockpit.

Here in Ritchie's own words is the story of Ritchie and DeBellevue's double kill on July 8: "By July 8 we had not flown for almost a week because of the weather; June had been a terrible month, with no victories for anyone at Udorn, and maybe only two in the theatre. The target on the 8th was a vehicle repair centre 30 miles south-west of Hanoi, and it was a fairly large, important spot. I was leading Egress Flight, the last flight of four and so named because you protected the first guys in, who were coming out low on fuel.

"Egress flights rarely got any action; most fights occurred in the first 10 or 15 minutes over the target, and I have to admit I was grumbling about the probable inactivity. Then we heard that the chaff aircraft, Brenda 01, had been hit in the left engine by an Atoll missile from a MiG. He was bleeding fuel and hydraulics and his fire-warning light was on. About the same time, Dallas 04 called in that he was hit and heading out. So there were two crippled planes coming out, altitude and heading known, a perfect set-up for the communist ground control to vector on.

"We were assigned to patrol some 60 miles north-west of Hanoi and were inbound there at the time. We immediately changed plans and headed for Brenda 01 and Dallas 04. I dropped from 15,000 to 5,000ft, for the MiGs had been popping up from low altitude, and proceeded to an area about 35 miles south-west of Hanoi. Sure enough, Red Crown and Disco both began calling out the positions of MiGs there.

"I was turning hard, never flying more than 6 to 8sec before turning at least 90°, feeling that this provided pretty good protection from an attack at 6 o'clock. Sure enough, we received the call 'Heads up', which meant that the MiGs had us in sight and were cleared to attack by their ground control. We were moving along at about 550kt indicated airspeed (KIAS), about Mach 0.96, just at the point that we could go supersonic if needed. I had been down and met the Disco people the month before, and had talked to them by secure landline that morning so that they knew our callsign, orbit point, nature of the mission and so on. Thus when the chips were down, instead of going through the standard prescribed transmission – 'Paula 01, this is Disco 23, you have bandits bearing 270° degrees at 2 miles, closing' – Disco simply said: 'Paula, they are two miles north.'

"I immediately turned north and picked up a MiG-21 at 10 o'clock level; if Disco hadn't called the timing was such that he could probably have fired an Atoll missile and may have hit us. That big old Connie, 150 miles away, saved the day. I rolled left, passing canopy to canopy with the MiG-21, less than 1,000ft apart. I blew the wing tanks, went full afterburner and slipped supersonic, reaching over 600 KIAS, just as he was doing. Imagine, a 1,200kt passing speed and we saw the guy in the cockpit. The MiG was a highly polished airplane, with bright stars; every one of the other 16 MiGs I saw in combat was a dingy grey. We knew there was another MiG about, and figured there was something special about this pair.

"I rolled level, went into a dive from about 5,000ft and waited, which was a very difficult thing to do, for he was either getting away or was reversing course to attack me. But from studying their recent tactics I knew there would be a second MiG in trail with the first if he was not in fighting wing formation. Sure enough, about 8,000ft behind was the second MiG; he missed us, probably because we were camouflaged and low against the jungle green. As we passed I went into a 135° nose-down slicing left turn, using max turn rate. About half-way through the turn I saw the No 2 MiG in a right turn, high, at 8 o'clock; it was a big mistake on his part: he should have been turning left. At this point we had a high track-crossing angle developing and would meet in the front quarter should the turn continue. I therefore barrel-rolled left, coming out at 5 o'clock, 6,000ft out and low, perfect for a Sparrow shot. He was high in the blue sky, an excellent contrast for the radar, which was already in boresight. I manoeuvred the target into the gunsight, hit the auto-acquisition switch on the left throttle and got an immediate lock-on, which is unusual. Often the radar would cycle two or three times before locking on.

"Chuck said: 'It's a good lock,' and it was then necessary to wait four long seconds before trigger squeeze. It took approximately 4 sec for radar settling and missile match-up. The MiG saw us and began a hard turn down into us, trying to disengage and break the lock. At the lock the MiG was about 5° angle off our nose, and after 4sec I squeezed the trigger twice.* By the time the first missile came off the MiG had out-turned us by about 40° and was about 45° off our nose. I was pulling about 4g. I had to turn tight enough to keep the radar lock on, but not so hard that the 3-4g launch limit for the Sparrow was exceeded. When the missile finally came off the airplane we were approaching the radar limit,

---

*Standard practice was to fire two missiles; hit rate at the time was about 16 per cent for Sparrows, so two missiles offered a greater chance of success. After the pilot had waited four seconds before trigger squeeze, the missile would sit for another 1½ agonising seconds while about 90-plus electrical and pneumatic sequences took place. The 5½sec seemed like forever in a dogfight, and the pilot inevitably thought that there had been a malfunction or misfire.

# CLASSIC HIGH YO-YO

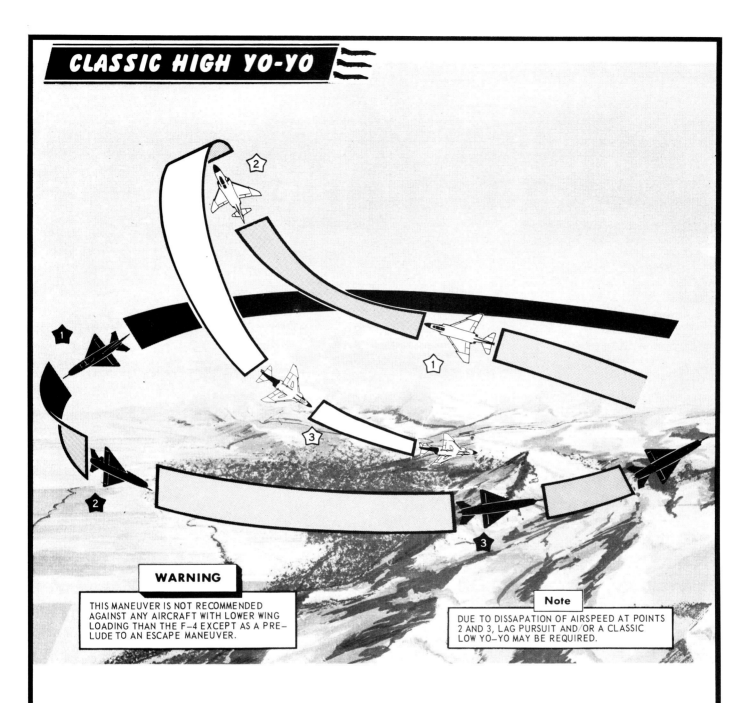

**WARNING**
THIS MANEUVER IS NOT RECOMMENDED AGAINST ANY AIRCRAFT WITH LOWER WING LOADING THAN THE F-4 EXCEPT AS A PRELUDE TO AN ESCAPE MANEUVER.

**Note**
DUE TO DISSAPATION OF AIRSPEED AT POINTS 2 AND 3, LAG PURSUIT AND/OR A CLASSIC LOW YO-YO MAY BE REQUIRED.

The high yo-yo is an offensive maneuver in which the F-4 maneuvers through the vertical and horizontal planes to prevent an overshoot in the plane of the enemy's turn. An overshoot can be generated by (1) too much angle off or (2) excessive overtake, or a combination of both. When the overshoot appears imminent, the F-4 should roll a quarter turn away and pull up into the vertical plane ①. This allows nose-tail separation to be maintained. Afterburner may be employed as required to maintain closure.

After starting the pull-up, the F-4 should keep the nose coming up and roll toward the enemy to keep him in sight. At the slower speed in the apex ②, the F-4 should pull his nose back down through the horizon to realign with the enemy's six o'clock position ③. The F-4 must avoid attaining an excessive nose-low attitude when realigning with the enemy's six o'clock. Recovery from this attitude is difficult and the enemy may force the F-4 into a defensive situation by pulling up into the attack.

**The classic high yo-yo. The F-4 exploits its performance in the vertical plane to get inside the MiG-21's turn.** *(US Navy)*

# BARREL ROLL ATTACK

In a situation in which the attacker approaches a defender at high angle-off and long range, the high yo-yo has questionable value. The purpose of the barrel roll attack is to reduce angle-off without loss of energy. It is possible to reduce turn and velocity by two methods: (1) through both the vertical and horizontal planes by employing a two-dimensional maneuver (yo-yo), or (2) mareuver through both the vertical and horizontal planes by employing a three-dimensional maneuver (barrel roll). Thus far, the emphasis has been on the first method. Now both methods will be employed to reduce angle-off. If the F-4 approaches the enemy from below and at a high angle-off, he continues to cut off, in an attempt to reduce angle-off, until he reaches his pull-up point. The F-4 should pull up well inside the enemy's defensive turn ①, then barrel roll in a direction away from the enemy's turn ②. If the enemy's defensive turn is toward the left, he should roll right; if the turn is toward the right, he should roll left. The roll is not a high-G-barrel roll. As the F-4 rolls nose-high through the inverted position, he should play back pressure and rudder to obtain a nose-low 270-degree change in direction ③. During this portion of the maneuver (from the inverted position to the 270° point of the roll), the F-4 receives the benefit of 1-G gravity, which assists him in gaining a rapid change of direction toward the enemy's six o'clock position. If the F-4 has made the entire maneuver (from pull-up through the roll) without an overshoot, he will be above, at a reduced angle-off, and in a position to dive below the enemy's line of flight for a six o'clock low position ④. LAG pursuit techniques can be used to retain the offensive.

**Barrel-roll attack, which reduces angle-off without loss of energy.** *(US Navy)*

the missile launch g limitation, and the minimum range limitation.

"The first missile guided perfectly and went through the centre of the MiG where the wings meet the fuselage and the airplane split in two. There was a large fireball, and the second missile went right through it; two perfect missiles under maximum-range launch conditions. The fireball was now in front of us; I pulled up hard to clear but a small piece of debris nicked the left wing.

"I expected the other MiG to leave the fight; the Soviets then did not seem to have a concept of mutual support, and in every other case in my experience, if one MiG got into trouble the other one split. But not the silver MiG. He stayed in the fight and tried to shoot down my No 4 man, Tommy Feazle, a young lieutenant with about 400hr flying time. Tommy called: 'Steve, I've got one on me.' I immediately unloaded [dumped g load] and started another drive for the ground, still in full afterburner. I didn't see the MiG, but I knew where he had to be from the geometry of the fight. I came hard to the right in a slicing turn, across the circle, into a 5 o'clock position behind the shiny MiG. The position was very similar to the first combat, except closer. He went into a very hard turn, rotating the highly manoeuvrable little fighter in mid-air. I put the MiG in my gunsight, hit the auto-acquisition switch, and Chuck confirmed a good lock. I counted 3.9sec (it takes a lot of discipline to do that, you want to squeeze right away). He was only about 3,000ft at about 60° angle off, the radar break-lock point. I was pulling 5g, beyond the capability of the missile.

"I already figured I'd missed him, and tried to reach the master arm switch to turn on the guns, but I couldn't make it because the g forces kept my hand from getting there. The missile launched. It's a big missile (we called it the '500lb bullet'), 12ft long, and it accelerates to 1,200mph above the launch speed in 2.3sec. The Sparrow has a 30lb proximity-fuzed warhead. The missile snaked out in front of the aircraft like a Sidewinder, and I was talking to it real hard: 'You son of a bitch, it's over *there*!' All of a sudden the missile seemed to do a 90° right turn and slammed into the MiG, which was itself pulling 7 or 8g trying to get out of the way. The airplane simply disintegrated.

"The only thing I had said in the first part of the fight was 'Splash One'. Now I called 'Splash Two'. The whole fight had lasted only 1min 29sec. In that time our fuel load had gone from 12,000lb to 7,500lb, which shows you how fast we used it in afterburner." The two kills were Ritchie's third and fourth victories, and DeBellevue's second and third.

Now here's how the fight looked to Chuck DeBellevue in the back seat of Paula 01: "On July 8, 1972, Paula Flight was patrolling south of Yen Bai, about 70nm west of Hanoi. We were the Egress CAP, the rearguard. Two aircraft in the Hanoi area, Brenda, an F-4, and Dallas, an F-105, announced that they had engine fires and were heading out. They were both talking a lot on the radios as they departed the Hanoi area. They wanted the rescue effort to get going in case they didn't make it. Of course the North Vietnamese heard this. We felt that the MiGs would try to intercept them and that it would be in everyone's best interest if we were to head east. We were advised by Disco and Red Crown, the two ground controlled intercept (GCI) stations providing us with information, that Blue Bandits (MiG-21s) were located approximately 35-40nm south-west of our position. We headed towards the threat in patrol formation at 4-5,000ft and crossed the Black River on a southerly course. We ended up in an area 35 miles south-west of Hanoi, under the control of our GCI. It turned out that our GCI was controlling us and the MiGs' GCI was controlling them. No problems there, because we had a good idea of where everybody was located. Situational awareness was very important. All of a sudden, a frantic call came from our controller: 'Paula, you're merged.' The Blue Bandits had merged with us: everybody was in the same spot on the controller's radarscope. For two minutes (it could have been an eternity) no-one saw any sign of the MiGs; not on the aircraft radar, not on the GCI scope, not visual. We had already got rid of the external fuel tanks and were running in afterburner to kill the F-4's telltale smoke trail.

"Although we were doing a lot of manoeuvring to ensure that no one got behind us unnoticed, I felt that the MiGs were somewhere in front of us. Suddenly, at 11 o'clock, just to the left of the nose, I saw a black fly speck against a white cloud. It didn't belong there. I called it out to Steve, who came left. A few seconds later there was a bright shiny new MiG-21 line-abreast with us and going the opposite direction. The MiG started a hard turn, not into us as we would have done, but away from us. He was expecting us to follow him. We started a hard slice to do just that but delayed it long enough for his wingman to show up. From flying every day we knew that they didn't normally fly single-ship. We didn't have long to delay: the wingman was following about 8,000ft in trail with his lead, and he also started a hard turn away from us. We already had the wings set for the slice and pulled into a hard 135° manoeuvre that placed us about 6,000ft in trail of the wingman. We obtained an automatic-acquisition radar lock on the wingman at about 5,000ft. He was in a hard right turn, pulling about 5g, when we fired two AIM-7 radar-guided missiles.

"We normally fired two missiles to increase our probability of kill (PK). The MiG was in a turn, so that the centre of the radar energy pointed at the area behind the canopy. The first missile hit the MiG-21 right behind the canopy. There was a large yellow fireball as the MiG broke in two and burned. The second AIM-7 never faulted but went through the fireball. The MiG continued to disintegrate until it impacted the ground. At that point we unloaded to get our energy back and to ensure that our flight was still in an offensive formation. Our No 4 called out that the lead MiG-21 was now on him. The lead MiG-21 had come full circle and was now a real threat. We came back hard into the fight again and obtained an automatic-acquisition lock on to the lead MiG. We were 4,000ft from the MiG-21. At this point the MiG driver started a hard, 5-6g, pull to get away from us. We launched the No 3 AIM-7 at 3,000ft from the MiG-21. This was maximum range for the missile, based on the parameters we had placed on it. A 5-6g target, about 70° angle-off (the lead MiG-21 had pulled off the radar scope so that there was no target on the radar for the missile to follow), 3,000ft range – the MiG-21 was on the edge of the envelope. After it is ejected from the aircraft the AIM-7 does a roll manoeuvre to align its antennae with the aircraft radar antenna. It then starts an initial turn as the missile motor fires, to get it heading in the right direction before aircraft guidance signals are received. This third missile did its initial manoeuvres underneath our F-4 and exited the wingtip area, heading right for the MiG. The missile impacted the MiG-21 right behind the canopy, resulting in a large yellow fireball. This MiG-21 also broke in two and began to disintegrate. The front of the aircraft was observed impacting the ground in a fireball. Paula Flight remained in tactical formation throughout the fight and egressed as a flight of four. GCI told us later that two additional MiG-21s that were being vectored in to help out their buddies were vectored away from us. It took about a minute and a half from tally-ho of the first MiG to the destruction of the second MiG."

The MiGs moved to higher altitude after that engagement. Surprisingly, the F-4 with a combat load is not very manoeuvrable above 25,000ft, and it was difficult to get at the lightly loaded MiGs.

Ritchie scored his fifth victory, becoming the first USAF ace of the war, while leading the MiGCAP force for a strike on the Thai Nguyen steel mill in Buick 01. He and DeBellevue were 43 miles north of Hanoi when DeBellevue picked up MiGs on the special equipment. The enemy was 37 miles south of Hanoi, almost 80 miles away. Buick 01 headed south, with DeBellevue getting a radar lock-on about 15 miles out in the front quarter. Ritchie did not fire the Sparrows head-on because of converging friendly aircraft. He performed a pop-up attack with a convert to the rear quarter, came in behind the MiG and launched all four Sparrows. The first two were fired out of range, hoping to make the MiG turn. The third missed, but the last one made Ritchie an ace.

Ritchie was then summoned for a series of debriefings, intended as much to protect him from possible harm as to derive intelligence benefit, and DeBellevue was assigned to fly with Capt John A. Madden Jr in Olds 01 on a MiGCAP mission. The targets were in the Thai Nguyen area and along the north-east railway. Flying with them in Olds 03 were Capt Brian Tibbett and 1st Lt Bud Hargrove. Madden would end the war with three victories, while Tibbett and Hargrove would score two each. Both Tibbett and Hargrove were killed later on a MiGCAP mission.

Olds Flight entered its assigned area 30nm north-north-west of Hanoi, in the vicinity of Thud Ridge, that rising line of karst which provided cover from the enemy radars. Allied GCI called a MiG-21 heading west, and directed Olds 01 to head east to Phuc Yen airfield. Olds 03 suddenly called out a SAM at 3 o'clock: it turned out to be a MiG-21 and Olds 01 turned to attack. The MiG was about 1,000ft above ground level, about to land at Phuc Yen airfield. DeBellevue achieved a radar lock-on and Madden fired two Sparrows, but the radar transferred lock from the MiG to the ground and the missiles missed it. Olds 01 started to over-run the MiG, so Madden bled off airspeed, dropped his flaps and slowed to about 180 KIAS, a fatally low speed to be stooging around North Vietnam. The MiG sucked up its gear and flaps and turned into the F-4, accelerating as it turned and pulling up and out of the engagement. Madden cleared Olds 03 to fire and Tibbett let loose two Sidewinders, which missed. He closed to gun range and hosed the MiG with 20mm from his gun pod. The MiG took hits and the pilot ejected.

Olds 01 called the element to join up and then responded to a request from Allied radar to cover the F-4s egressing from the target area. Chevy 01 and 02 were coming out on emergency fuel. As Olds crossed the Red River, West of Hanoi, DeBellevue achieved a radar contact on two MiGs closing to engage. DeBellevue called out Olds 01's fuel status: they had to make short work of the MiGs or they would not have enough fuel to get home. Then Olds 03 called out: "Hey John, they're 19s," which meant they had a highly manoeuvrable tiger by the tail.

Olds 01 went into a hard, 8g turn while Olds 03 pulled up to act as cover. 01 was about 90° through his turn when the MiGs jettisoned their external fuel tanks and began turning hard into him. Olds 01 fired two AIM-9J Sidewinders at the trailing MiG, painted dark blue. The first missile detonated near the left wing and stabiliser of the MiG and it left the fight, burning. Olds 01 then engaged the first MiG, which was painted emerald green. After a very hard turn the next AIM-9 produced a loud growl in the crew's headsets, indicating that it had picked up the heat of the MiG's engine. They fired the third Sidewinder, which impacted in the MiG's afterburner.

Olds Flight then left the battle area amidst intense anti-aircraft fire that included everything from 23mm to 85mm. They were 20 miles south-west of Hanoi and Olds 04 called to say that he had about 1,800lb of fuel, some 6,000lb below bingo. Olds 01 directed 04 to climb to altitude, and fell back to protect him from a number of MiGs which had been sighted by Allied radar. Olds 04 flamed out about 150 miles north-east of Udorn, their home base, and the crew ejected. They were picked up by a helicopter about 20min later. The rest of Olds Flight went straight back to Udorn. The back-seater in 01, Chuck DeBellevue, had just become the leading ace of the Vietnamese War.

The final twist in the aces controversy was the case of Captain Jeffrey S. Feinstein, a weapon systems officer from East Troy, Wisconsin. Feinstein scored his five victories with no less than four aircraft commanders between April 16 and October 13, 1972. Those aircraft commanders scored only with Feinstein aboard, one sharing in two victories with him, the others collecting one each.

Feinstein's third and fourth victories were obtained with Lt-Col Carl G. Baily in the front seat of their 13th TFS, 432nd TRW, F-4D. They were proceeding to their assigned orbit point near Kep airfield on July 29 when Red Crown called out the MiGs. After losing the initial contact Feinstein locked on again and Baily fired off his Sparrows, destroying the MiG. Said Baily afterwards: "The MiGs were coming at us at very high speed; they managed to get by us before we engaged them. We turned as hard as we could, started towards them and got them right in front of us, coming head-on. Jeff locked on to the MiG and I fired two missiles. They both guided right in and splashed him good. The credit all goes to Jeff; when you get them head-on, the guy with the radar does all the work. I just sat up front and squeezed the trigger."

The question of one man or two in the cockpit is still live. The F-14 still requires two while the F-15, F-16 and F-18 do not, although there are variants of these types that have two seats. Equipment has become more automated, but then more equipment has been installed. Missions have become more specialised and demanding, even while aerodynamics have been significantly improved. It really doesn't matter much, however: the F-4 will be soldiering on through to the end of the century, and it will be doing it with two good men in the cockpit.

An Egyptian Air Force F-4E; fortunately there has never been an occasion for a mutual Israeli/Egyptian F-4 combat. (Courtesy Ted Van Geffen)

The F-4E was exported in larger numbers than any other Free World fighter. (Courtesy Ted Van Geffen)

Operating the F-4E in the hot desert climate of the Middle East posed maintenance problems far beyond those originally conceived of by McDonnell Douglas. (Courtesy Ted Van Geffen)

An RAF Phantom FGR, September 1972; the difference in appearance occasioned by the installation of Rolls Royce engines is minimal.

An F-4J of VMFA 321, at Washington D.C. in August, 1975. The Marines relished having a first line fighter.

Da Nang, in what used to be the Republic of South Vietnam. A Marine RF-4C rolls past quonset-hut style hangars – one wonders what they house now. (Via James Gatewood)

In 1971, VMFA 115 was operating F-4Bs out of Da Nang. (Via James Gatewood)

Ah, the colors and the markings – so much more satisfying than today's low visibility requirements. An F-4D from VMA 115 at Da Nang in 1971. (Via James Gatewood)

The F-4J's were a great improvement when they arrived; these are at NAS Oceana.

Sitting broiling in the hot sun, the Phantom's cockpit rose to blazing temperatures – and one should never forget the hard working mechanics who burned their hands on tools keeping the planes in service. (Via James Gatewood)

No one could mistake the brightly striped tail of VF 151's F-4B's. (Via James Gatewood)

Da Nang was one of the world's largest supply depot's; after this F-4B of VF 161 took off, the crew could glance down at a billion dollars worth of equipment stored over the vast acreage. (Via James Gatewood)

**This F-4B of VF 151 just flew in from the U.S.S. Midway. (Via James Gatewood)**

**You could keep the forward section of the F-4D sparkling clean, as this one from VF 161 shows, but there was nothing you could do about the exhaust streaks from those roaring J79s on the aft end. (Via James Gatewood)**

A scene many an insolvent Philippine businessman would like to see again: U.S. aircraft stationed at Cubi Point. Here an F-4B, an F-8, and an A-7 await their next flight. (Via James Gatewood)

The men who maintained the Phantoms were highly respected by the flight crews, and deservedly so. Highly skilled work at low pay under bad conditions, and Jane Fonda at home complaining.

Refueling was done often, but the hazard implicit in so much metal moving so fast, so close together, prevented it becoming routine.

The Phantom had more angles than any fighter before or since; despite them it was a beautiful bird.

**What a line up!! From the right: Convair F-106A, McDonnell Douglas F-4C, McDonnell Douglas F-101B, and a Lockheed F-104. This was before the days of declining budgets.**

**Somewhat awkward appearing on the ground, in the air, with gear and flaps up, the Phantom took on a sleek, lethal look. (Courtesy William Vasser)**

**Clark Field before the volcano came and the Americans left. An F-4D of the 46 TFS taxies by. (Via James Gatewood)**

**A friend in need** – the KC-135 crews were vastly appreciated not only by Phantom pilots, but by everyone doing aerial refueling in Southeast Asia. The tankers took risks far beyond any they had been trained for, entering the combat zone, and vulnerable to MiG strikes.

**There's always time for a little hot-dogging, even over South Vietnam.**

An F-4D of the 7 TFS basking in the glow of the Clark Field afternoon sun. (Via James Gatewood)

The insignia showing two MiG kills distinguish this F-4E of the 469 TFS, 388 TFW at Andersen AFB, Guam, in February 1973.

A line up of F-4Es at Andersen AFB, Guam.

Udorn Royal Thai Air Force Base, Thailand; an F-4D taxies past a Lockheed C-130 Hercules. (Via James Gatewood)

An F-4E taking off at Da Nang, 1971; the scene doesn't convey the immense amount of traffic that Da Nang handled day in and day out. (Via James Gatewood)

The F-4Cs of the 557 TFS. (Via Bill Reeves)

These are the moments of quiet beauty that makes flying so addictive.

The delicate moment before touchdown; an F-4E of the 469 TFS landing.

The whole purpose of the Phantom is captured in the two red stars indicating MiG kills; an F-4E of of the 469 TFS at Andersen AFB, Guam.

**The work of the maintenance men was never finished; the armament troops were especially overladen with work and responsibility. (Courtesy William Vasser)**

**The glorious moment of the hosedown after the 100th mission North; this is the 480 TFS.**

Takhli Royal Thai Air Force Base, Thailand; an F-4E readies for the next mission.

Colors and markings have changed over the years, but even camouflage couldn't mar the lines of the Phantom. This F-4D is from the 49 TFW.

Sleek and quiet, but not quite ready for the next day's work. (Via William Vasser)

An F-4C with unusual canisters, possibly carrying sensors for Igloo White.

The business end of a MiG-killing Phantom.

# 6. Naval and Marine Phantoms in Vietnam

The US Navy's and Marine Corps' use of the Phantom in South-east Asia paralleled in many ways that of the USAF. They were equally constrained by the rules of engagement, and funding and training problems had resulted in the slack of deficient planning being taken up from the lives and careers of the personnel. But everyone in the Navy and Marines felt that the war could be won, if only common sense was allowed to prevail. Other frustrations resulted from such misapplication of resources as the risking of a flight of F-4s against a wooden bridge over a narrow stream that could be replaced in an hour by native workers.

The *raison d'être* of an expensive carrier force is its ability to project power rapidly to almost any point in the world. US carriers have been pre-eminent in the Pacific since Japan's eclipse in the Second World War, and have distinguished themselves in both peace and war. When the Vietnam War seemed imminent the Navy was ready, with USS *Ticonderoga* and *Constellation* making the initial strikes against North Vietnamese bases on August 6, 1964. It was a small beginning to an agonising eleven years which reached their humiliating denouement in April 1975, when USS *Midway, Enterprise, Coral Sea* and *Hancock* covered the final escape of South Vietnamese soldiers and officials and their families as the North Vietnamese tanks rolled into Saigon. There was a brief further flurry of involvement when *Coral Sea* helped to resolve the *Mayaguez* incident.

Over this decade the US Navy maintained a mighty presence off the coast of Vietnam, flying untold numbers of missions and dropping millions of pounds of bombs. Some naval fliers devoted half a career to the endeavour, and through it all the McDonnell Douglas Phantom performed all of its many roles.

Phantoms scored the first MiG kills of the war on June 17, 1965. Two F-4Bs of VF-21, operating from *Midway*, were vectored to engage a flight of four MiG-17s near Hanoi. In a classic head-on engagement in which the closing speeds approached 1,100mph, the radar-guided Sparrow proved its superiority over the gun-equipped MiGs. Cdr Louis C. Page (executive officer of VF-21) and his RIO, Lt Jack E. Batson, got one, and the No 2 aircraft, flown by Lt John C. Smith Jr and RIO Lt-Cdr R.B. Doremus, got another. The pattern was set.

During the ensuing years 61 more MiGs would fall to Navy guns and missiles, though not all to Phantoms, of course. This works out at something like one every two months and clearly

**Below: An F-4B from VF-201 being prepared for launch. Asymmetric loading of centreline and port wing tanks did not prove to be a problem.**

underlines the fact that while the MiG war got the headlines, the bulk of the Navy effort was devoted to the same sort of thankless ground attack operation as was being carried out by the USAF. The following account by then Capt, now Lt-Col, Lawrence G. Karsch of Marine Fighter Attack Squadron 542 describes one of the high points in an unrewarding campaign.

Karsch had been in Vietnam for about six months and had acquired razor-sharp proficiency in delivering air-to-ground ordnance. On August 7, 1972, Karsch took off on a clear, hot, muggy day, like so many in Vietnam. Far from worrying about the mission, he was relieved to be doing something away from the boredom of the alert pad at Da Nang. Here is his story: "My flight leader, Maj Fritz Menning, and I and our two RIOs, all of Marine Fighter Attack Squadron 542, pulled the afternoon air-ground hot pad at Da Nang on August 7, 1969. Our two F-4Bs were loaded with napalm, Snakeye bombs and 2.75in rockets. Maj Menning's aircraft had the Snakeyes and rockets and I had the napalm: seven 500lb fire bombs.

"I liked the air-ground hot pad because it got a lot of business and I knew that when we were scrambled it would be for something important like troops in contact. We didn't have to wait long for a mission. No sooner had we assumed the hot pad than the klaxon sounded. Maj Menning and I sprinted from the wooden shack we called a ready room towards the aircraft. Marines on the flight line were already at the aircraft waiting for us. Our two RIOs, Capts Al Graf and Flip Flanigan, joined us shortly after getting the mission particulars over the landline.

"We got off from Da Nang in good time and headed north. En route, Al Graf told me that there was a bad situation on the DMZ and we needed to check in quickly with a Marine OV-10 airborne FAC with the callsign Hostage. When we contacted Hostage he told us that a Marine reconnaissance patrol had run into a large NVA force and were about to be overrun. The Marines were on the top of a small hill and about two companies of NVA had advanced to within 50m of their position. Hostage told us that the patrol needed our ordnance immediately, but only napalm could be used at first because the NVA were too close to the Marine position to use bombs or rockets. He went on to say that we had to put the napalm within 50m of the Marine position.

**Right: Deck crewmen have to be agile and constantly on the alert. Lack of concentration can be fatal. This crewman has just finished fitting the catapult bridle to the F-4B and is sprinting clear as the flight deck officer signals "all set" to the pilot.**

**Below: Phantom laden with rocket pods and centreline tank launches from the waist catapult.**

"That transmission practically made my heart stop. As I had all the napalm and because the napalm canisters tumbled through the air and thus were not particularly accurate, I was suddenly faced with the possibilities of either not putting the napalm on the NVA or incinerating our people. Hostage cleared me in hot following a dry run by Maj Menning and said to put my napalm 30m down the hill from an air panel the Marines had spread out to mark their position.

"With the biggest lump imaginable in my throat and a word of encouragement from Al Graf, I let two nape cans go and just prayed that I hadn't hit the Marines. After what seemed like an eternity Hostage shouted: 'Great hit, Ringneck! Right where I wanted it!' What a relief! What a sense of accomplishment!

"What followed was a rout. We progressively put more napalm, Snakeyes and rockets on the retreating NVA. A few days later, when a Marine battalion swept through the area, they counted 54 NVA dead and a larger number of weapons. We had badly punished a large NVA force that had been on the verge of overrunning the Marine patrol.

"There were two other missions that afternoon off the hot pad, but I don't remember them. I do recall that both Al Graf and Flip Flanigan were killed a few weeks later. But I shall never forget that day on the DMZ when Marines in the air were there for support when Marines on the ground needed it in the worst possible way."

Things were not always so well organised. Rear-Adm Harry Gerhard USN (Retd) recalls that in the early days the Navy used what was essentially a combination of Second World War and Korean War tactics, tentatively modified for high-performance jets. Gerhard notes that early in the war one school of thought believed that jet attacks could not be made at dive angles greater than 30-35°, a doctrine totally disproved by the F-4's 45° dive capability.

But F-4 operations cannot be considered out of their full context. Far more than the Air Force, the Navy operated as an organic unit in which the carriers and their multi-type air wings were harmonised into an attack force which could operate continuously, around the clock, day in and day out. The cost of this effort was enormous. In the early days of the war, for example, there were three to five aircraft carriers in

**Left: It takes a lot to get a big jet fighter on to a carrier deck: Phantom comes aboard with gear, hook, flaps and leading-edge droop all deployed.**

**Below: Smoke streams from the tyres and sweat from the crew as a Phantom thuds on to the deck. The hook has safely taken a wire but for some reason the catapult launch bridle seems still to be attached to the aircraft.**

Above: The lovely but deceptive Vietnamese dawn. A Phantom of VF-142 ("Ghostriders") awaits developments. These markings were current in 1965. *(William C. Vasser)*

the theatre at any one time; later, in 1971-72, five was the norm. One would be on Dixie Station (about 70nm due east of Cam Ranh Bay); one would be at Yankee Station (70 to 100nm out in the Gulf, about due east of Vinh); and perhaps one would be off the North, anywhere from east of Thanh Hoa all the way up to Haiphong. In the northern waters the carriers were always at least 70nm offshore to ease the MiG threat. Then there would be a carrier in transit to port (Subic Bay in the Philippines, Yokosuka in Japan or, if they got lucky, Hong Kong). A cruise was typically nine to ten months long, with a six-month turnaround in the US for refit and training.

Phantoms operated off *Midway* and the larger carriers, with a basic air wing consisting of two squadrons of F-4Bs, two or three squadrons of A-4s, one squadron of A-6 attack aircraft, and a unit of E-2B Hawkeye early-warning aircraft. (Later in the war the A-4s were gradually replaced by A-7s.) On the supercarriers the air wing would be expanded to include two squadrons of A-7 Corsairs, four or five KA-6D tankers, and a reconnaissance squadron of six RA-5 Vigilantes. There would also be an assortment of helicopters, ECM aircraft and so on.

Early in the war the Navy aircraft did not have any radar warning or ECM devices. The earliest, little more than an improvisation, was called "Little Ears" or "Big Ears," depending upon your source. It consisted of a box about the size of a cigarette packet which was equipped with suction cups and yards of wire. Each aircrewman would check out a Little Ears in the ready room and carry it to the aircraft. The box would be attached by suction cups to the inside of the canopy, and a connector on the end of the wire plugged into the communications line just below the crewman's oxygen mask. The box would detect incoming radar signals and convert them to raw audio for the aircrew to hear. It was not very discriminating, so that any time the aircraft was in radar range of North Vietnam the crew would be subjected to a strange cacophony of noises resembling the sound of feeding time at the zoo. Many hours were spent in training sessions with a tape recorder, trying to learn how to recognise by sound various types of radars in each of their operating modes. Most aircrewmen became quite proficient at the task, for it was a matter of survival. Fortunately, by 1968 much more sophisticated and less demanding systems were available.

The efforts of the carriers were co-ordinated by the commander of Carrier Task Force 77, which was usually on duty on Yankee Station. Each carrier conducted flight operations for 12hr, followed by 12hr of maintenance; this method was known as "cyclic ops". During the 12hr of flying the normal routine was to launch a group of aircraft, or cycle, every 90min. On the noon-to-midnight schedule the first cycle launched at noon, the second at 1330hr (the first-cycle aircraft recovering immediately after the second launch was complete), and so on. An average daytime launch would consists of six F-4s, six to eight A-6s or A-7s, two to four A-6s, two to three tankers, one Hawkeye, one Vigilante and one ECM aircraft. A "planeguard" rescue helicopter was airborne for all launches and recoveries.

One carrier would fly from noon to midnight, then another would take over and fly from midnight till noon. If there was a third carrier available it would overlap, generally from 0600hr to 1800hr. It was however often assigned to fly "Alpha strikes" against selected targets in North Vietnam. An Alpha strike would typically consist of 40-45 aircraft. Eight to ten fighters would serve as CAP or escort; 12-14 light attack and four to six medium attack aircraft would carry the bulk of the ordnance. Two Hawkeyes would be used to assist with vectors to the coast-in points, and to keep track of those safely "feet wet" (past the shoreline) on the return flight. KA-6Ds and A-7s with D-704 buddy refuelling stores would be launched, as would one or two Vigilantes for bomb damage assessment. From two to six A-7s armed with Shrike and one or two A-6s carrying Standard Arm anti-radiation missiles would be launched against SAM or AAA sites. By 1972, when the decline in the ground forces was demanding extra effort from the air, three carriers would sometimes conduct Alpha Strikes in concert for several days in a row. These were co-ordinated as closely as possible to make optimum use of the MiGCAP and reconnaissance aircraft.

Each carrier deployment, usually lasting nine or ten months, would be divided into five or six line periods, each normally of 35-45 days but occasionally extending to 60, 90 or even longer if things were hot. After a line period the carrier would get 10-14 days off the line, of which perhaps eight or nine would be in port. While on the line each carrier would also stand down for one day out of every six to twelve.

Dixie Station aircraft worked in support of operations south of the DMZ. A carrier would normally spend at least

*Above:* The F4H-1's carrying capacity was underestimated from the start. The weights and combinations of weapons achieved in Vietnam surprised even McDonnell engineers. *(MDC)*

*Below:* Lt-Cdr Gene Tucker and LtJG Bruce Edens pose aboard USS *Saratoga* with their newly applied MiG kill marker. This one is more explicit than most, showing the Sparrow knocking off the MiG, all superimposed on a red star. *(US Navy)*

*Bottom:* MiG kill emblem on Phantom from VF-33, embarked in USS *Independence*, indicates that the victim was a MiG-21. "E" is Navy's "E for Excellence" award. *(William C. Vasser)*

part of its first line period working in a lower-threat environment so that new pilots could be taught the ropes and the bugs worked out of procedures and tactics. All of the fighters and attack aircraft would carry Mk 80-series bombs and rockets for use in support of ground operations, working under the control of a FAC or Fast FAC. On rare occasions the Navy aircraft would join a Loran-equipped USAF F-4 for a multi-aircraft bomb release, usually in overcast weather when normal target acquisition was not possible.

Navy F-4 missions were generally similar to their USAF counterparts, most being ground support strikes. The Navy would receive a "frag" message daily from 7th Air Force Headquarters in Saigon, and targets would be assigned to the various carriers by the commander of Carrier Task Force 77. Small targets such as a group of WBLCs (waterborne logistics craft: sampans, etc) or a truck park would be assigned to an individual squadron. These missions were generally flown in a two-aircraft section, although occasionally a division of four F-4s would work a target together. Occasionally, as the Navy F-4Bs and F-4Js did not have inertial navigation systems, several sections whose targets were close together would go in together for mutual navigational support, and to avoid the possibility of collision.

F-4 weapon loads on strike missions varied according to target composition and ordnance availability. Cluster bombs (CBUs), especially prized for use against trucks, armoured vehicles and flak emplacements, were frequently in short supply. The F-4s always carried a mixed bag of missiles for air-to-air, usually two AIM-7E Sparrows on the aft fuselage stations and two to four AIM-9 Sidewinders on wing stations 2 and/or 8. The Navy never acquired an internally mounted gun, and the Mk 4 20mm gun pods were considered unreliable. Moreover, they could only be carried on the centreline station, No 5, where Navy pilots preferred to carry their external fuel.

Triple ejector bomb racks (TERs) would be carried on stations 1 and 9 (outboard) or 2 and 8 (inboard), with the most common bomb load being six Mk 82 low-drag general-purpose (LDGP) bombs with conical fins. Sometimes Snakeye fins (retarding type) were used. Only rarely were Mk 83 LDGP 1,000lb bombs carried.

South of the DMZ napalm was often carried, especially early in the war. The Mk 80-series bombs were frequently fuzed in both nose and tail. A mechanical fuze pre-set with the arming and detonation delays best for the target in

Above: Pictured after the war, at Barksdale AFB, Louisiana, this F-4B from VF-103 still sports its MiG kill marking.

question was set in the nose. The tail fuze was electrical, and the detonation delay could be set in flight by the pilot if the mission was changed or an alternative target designated.

When a target was assigned the flight leader would assemble his flight to plan the mission. Each carrier had an intelligence centre with as many as eight or ten intelligence officers. These specialists would provide charts, photographs, target data and details of current enemy anti-aircraft sites and MiG order of battle. Ingress and egress routes and ordnance delivery limits would be decided, and each crewmember would mark the intended routes on his chart.

The typical Navy F-4 strike mission was characterised by 30-45min of high activity, followed by 45min of boring "max conserve" holding before recovering aboard the carrier. Immediately after launch the F-4s would join up with a tanker (usually overhead but often en route to the target) and take on 2,000lb of fuel. The flight would then proceed to the target and drop the ordnance as soon as possible because the high drag of bombs or pods caused fuel consumption to soar.

The F-4s virtually always flew in "combat spread" formation. The wingman would be within 10° of abeam at 1½nm distance, and stepped up 1,500-2,500ft. They would fly to the coast at about 15,000ft and 300 KIAS, accelerating to 400 KIAS when five to ten miles from the coast. At the target they would carry out a high-angle dive (40-45°) with roll-in from about 11,000ft, releasing at 5-6,000ft and 500 KIAS. Most crews dropped their bombs in the ripple mode, but some would pickle off three quick pairs. When a section attacked, only one aircraft at a time would make a bomb run; the other crew would watch for flak, keep an eye out for MiGs, and observe the results. The leader usually rolled in first, with his wingman rolling in on a heading anywhere from 45° to 180° off that used by the leader.

Bombs gone, the section would then head for the Gulf, slowly climbing, jinking and holding 400 KIAS or better. Above 3,500ft they were safe from small-arms fire. After "coasting out" the section would slow to about 250 KIAS, join up in cruise formation, check each other over for any damage and then climb to 18-20,000ft to hold near the

Below: The Phantom had to take it as well as dish it out. Flak did this to an F-4, striking the underside of the aircraft just forward of the starboard jetpipe and severely damaging the engine. *(MDC)*

carrier at maximum-endurance speed until it was time to recover. (Even at max endurance, fuel consumption was about 100lb per minute.)

Capt Gene Tucker, who supplied this view of naval operations, believes that 70 per cent or more of the naval aviators who were shot down violated one or more of five basic combat precepts. These were:

1 Don't get low or slow.
2 Stay above 3,500ft altitude.
3 Don't make more than one run on the target.
4 Subsequent aircraft should not fly down lead's flightpath.
5 Jink. Jink. Jink.

On the Alpha strikes F-4s would be assigned as strike aircraft, TarCAP (target CAP) or MiGCAP. The only difference between strike and TarCAP aircraft was that after bomb release the TarCAP aircraft would remain in the target vicinity until the last aircraft had released its ordnance. TarCAP would then follow the strike group out. F-4s would also frequently be used as flak suppressors, proceeding to the target with the strike group but accelerating ahead when one or two minutes out and dropping flak-suppressing ordnance such as CBUs, Rockeye fragmentation bombs, or variable time-fuzed Mk 83 bombs. They would hit the active or known AAA sites 10-30sec before the strike aircraft rolled in.

Later in the war, in 1972, the F-4s added laser-guided bomb (LGB) and mining operations to their repertoire. The LGB, a normal Mk 82, 83 or 84 bomb fitted with a laser seeker head and guidance fins, required the use of a hand-held laser designator which looked not unlike a Brownie camera. The designator aircraft would proceed to the target with the strike aircraft (normally a section of A-7s but sometimes F-4s carrying the LGBs). The strike aircraft would drop their ordnance while the designator aircraft flew a steady semi-circle around the target so that the RIO could illuminate the target with the hand-held designator. Accuracy was good but this system was not healthy in a high-threat environment like Hanoi because the designator aircraft's flightpath was predictable, and because both the pilot and the RIO were "padlocked" (paying 100 per cent visual attention to the target).

The F-4s also dropped Mk 82 bombs configured as mines, with Snakeye fins and influence fuzes. Since the early Navy F-4s, lacking INS, could not navigate with the precision needed for the laying of an accurate minefield, they always flew on the wing of an A-6 during mining operations. The formation would fly the whole mission at low altitudes, from 200 to 400ft, to ensure accuracy and reduce exposure to SAMs and AAA.

The Phantoms also escorted the RA-5C Vigilante photoreconnaissance aircraft. There would be a "photo bird" airborne on every cycle, getting updated target photos, checking out road segments, or performing bomb damage assessments (BDA). Each would have an escort. Photo runs required large amounts of fuel, and the F-4 would get a big shot after launch (5,000lb or more), and would sometimes need another 2,000lb on coasting out. The RA-5C flight profile varied, but it was typically Mach 1-1.2 at 3,500-5,000ft, a speed range at which the F-4 is very fuel-inefficient. The F-4 had problems occasionally keeping up

Above: The Phantom proved rugged in Vietnam: it could take a near-miss readily, and was even able to absorb direct hits from the larger AAA. Redundant structure prevented damage as severe as this from leading to catastrophe. *(MDC)*

with the Vigilante because the big North American type could accelerate so rapidly. Photo escorts generally positioned themselves 1nm abeam of the recce aircraft, stepped up and positioned towards the most significant threat. Escorts reported flak or SAMs which threatened the photo aircraft because the view from the Vigilante was very limited.

Favourite mission of all was naturally MiGCAP. During the early years of the war, 1968 and before, the carriers would position a section of F-4s as MiGCAP just off the coast, or have them transit North Vietnam at somewhere around latitude 18° 30′. Later, especially in 1972, MiGCAP was used only in connection with major Alpha strikes or in conjunction with the B-52 raids. At least two, and sometimes four, sections of MiGCAP were assigned to cover each Alpha strike. Each section would be assigned a station over North Vietnam, normally located between the target and the closest known MiG threat.

MiGCAP aircraft would depart early in the Alpha strike

Marine RF-4B. Essentially similar to the USAF's RF-4C, this aircraft gave remarkable service in Vietnam. *(MDC)*

launch. Each section usually had its own KA-6D tanker. The section would rendezvous with the tanker and proceed towards the coast-in point. The flight leader would have computed the exact time to coast in so as to be in a position to provide protection as the strike group proceeded towards the target, and to be at the assigned MiGCAP station a few minutes before the strike was on target. Keeping this in mind, he would have to decide when to start refuelling his fighters so that they could get as much of their assigned fuel as possible (usually 5,000-6,000lb each) and be completed by the time they were 10nm offshore.

The fighter section would coast in at 400 KIAS or better, and 10-15,000ft altitude. The section's mission was not to jump any MiGs they could get their hands on, but to intercept any that posed a threat to the strike group. This was sometimes hard for fighter leads to keep in mind, as their pilots wanted nothing more than to get into air-to-air combat.

One of the biggest difficulties with MiGCAP was communications, given that there was always a high threat from AAA, SAMs and MiGs. The CAP frequency was different from the strike group's frequency, and the CAP had to depend upon the controller to pass on information about the progress of the strike group. Sometimes the controller would become engrossed with other problems, with the result that the strike group would have hit their targets and departed the area five or ten minutes before the MiGCAP was notified.

During 1973 Navy A-6s were carrying out single-aircraft low-level night strikes all over the area between Hanoi and Haiphong. MiGs were known to launch at night, and so a single MiGCAP F-4 was positioned along the coast at night. If a MiG was launched anywhere near the A-6s the F-4 would be vectored towards it; the enemy aircraft would invariably break off when the F-4 got within 25 miles or so. The following is an account of a night MiG kill by Lt-Cdr (now Capt) Gene Tucker, who provided much of the material for this chapter: "I was deployed to the Western Pacific as operations officer of the 'Sluggers,' Fighter Squadron 103 (VF-103). The squadron, at that time commanded by Cdr Bob Cowles, was part of Carrier Air Wing 3, whose air wing commander (CAG) was one of the top tacticians of the period, Cdr Deke Bordone. We were operating from the deck of the carrier USS *Saratoga* (CV-60), commanded by then Capt 'Sandy' Sanderson.

"The two fighter squadrons each maintained a continuous 5min alert F-4J Phantom and crew on the flight deck during periods specified by CTF-77. This duty rotated among the carriers on Yankee Station, usually in 12-24hr periods.

"On August 10, 1972, my RIO, Lt JG Bruce Edens, and I were standing the 1800-2000hr 'Alert 5' for VF-103. A fortunate quirk of fate caused us to have the alert at that time. The squadron executive officer, Cdr Danny Michaels, was scheduled to have that particular alert, but he had an important meeting to attend so he had swapped watches with me. A crew from our sister squadron, VF-31 'Tomcatters,' was also standing the Alert 5.

"The first indication which Bruce and I had that anything was going on was a loud 'Launch the Alert Five' over the 5MC (flight deck PA system) and 1MC (ship's PA system) at just about sunset. At the time I was sitting on the starter tractor beside the plane drafting proposed changes to the squadron standard operating procedures. Bruce was in the back seat, strapped in as we were supposed to be, and started yelling at me to get in and get the bird started. Ever since early in my first Phantom squadron tour I have done a 'scramble' start each and every time, so I was well prepared for the simultaneous strap in and fast start which followed. I

was quickly ready to taxi from my spot just aft of the island to the catapult. Because of the requirement to be able to launch the Alert 5 fighters immediately from any catapult in any sea state or natural wind condition, we were not loaded with a full bag of gas. As I was spotted on the catapult, I was ordered to take on a full bag of fuel, since there was no longer an active contact, and the ship was making plenty of wind to launch me. I did not know why at the time, but later found out that the MiG which had been tracked from the vicinity of Kep on a southern track had disappeared from the scopes in the vicinity of Vinh. My configuration was two AIM-7Es loaded on the aft fuselage stations, two AIM-9Ds on the wing stations (I don't remember whether No 2 or 8), a centreline fuel tank, and TERs on the outboard wing stations, 1 and 9. (We F-4s were doing lots of bombing, including laser-guided bomb designating, in addition to the MiGCAP, BarCAP and TarCAP missions in those days so we frequently left the aircraft configured for bombing missions by leaving the TERs on the wings.)

"As I completed fuelling the MiG reappeared in the vicinity of Vinh, and I was given a vector west and launched. I stayed right on the deck as I made a hard starboard turn off the cat to a westerly heading, accelerating to 450kt or so, then starting a climb. I had launched shortly after sunset. I was given one or two bogey calls – he was west of me at some 70nm – and then he disappeared from the controller's scope again. Bruce and I were disappointed as we were told to max endure and take up a CAP station at 15,000ft altitude in the vicinity of Hon Mat Island just off the Son Ca River mouth, east of Vinh. The VF-31 F-4 launched shortly after we did, followed by a VA-75 KA-6D tanker. The tanker reached our vicinity and Bruce and I joined him and commenced topping off just in case. It was completely dark by now.

"We had almost completed tanking when our controller asked which F-4 was on the tanker. I suspected why he was asking – a bogey contact – and since we were just about topped off I quickly unplugged. Bruce responded, before I could get unplugged, that we were, and the controller gave the VF-31 F-4 a westerly vector with information on a bogey some 10-12 miles west of him. He was also advised not to go feet dry without a radar contact. I immediately turned west and estimated that I was about 5nm north of the VF-31 F-4. We were both fairly close (5nm or so) to the coast, and he reported no radar contact and turned right to stay feet wet. I immediately told the VF-31 F-4 to reverse to the left as he turned east so he would be turning away from me and would not interfere. I then asked for 'bogey dope' (position of the bogey from me) and bogey altitude. We were told he was at 8,000ft and that he was about 140° at 12nm from us.

"I descended immediately to 8,000ft and Bruce got a radar contact south-west at 12nm almost at the same time. The bogey was tracking north and we rolled in about 8nm behind him. I plugged in the afterburner, accelerated to about 650kt, and we had closed to about 4-5nm when we lost radar contact. At that time we were closing fairly rapidly and, not wanting to overrun him, I slowed to about 400kt while doing a left 90° turn, then back immediately to the northerly heading. We realised that he had probably descended, so I let down immediately to 3,500ft. (I had made a previous Vietnam cruise and was familiar with the area north of Vinh and remembered that the highest karst mountains in the vicinity were not over 3,500ft.) As soon as we reached 3,500ft Bruce reacquired the bogey about 6-7nm in front of us. I remember saying to myself: 'Tucker, this is probably your last chance to get this guy, or any MiG, so you'd better make it quick, and good.' As soon as Bruce got the radar contact I had re-lit the afterburners and had accelerated to somewhere between 550 and 650kt, when I decided I should jettison the

**Two Phantoms from VMFA-531 take off. The squadron emblem, a skull with lightning streaming from the eye socket, became famous in Vietnam.**

centreline tank and TERs. My airspeed was much higher than the published jettison limits, but I applied about +2g and jettisoned the partially full centreline tank. There was a mild thump as it cleanly departed the airplane. (Due to the speed, there was already a moderate buffet.) I then jettisoned the TERs, applying about 1.5g or less, and was surprised at the pronounced 'bang' or 'thunk' on the wings. (I shouldn't have been: even though TERs are fairly light, they were well out on the wings, and even at small positive g developed a considerable force through the lengthy moment arm.) Now with a clean F-4, I was moving at 650-750kt. I don't remember exactly, but we were closing him at some 300kt even though we were directly behind him.

"We closed to about 3nm and were inside max range for our Sparrows. Bruce said 'Shoot, Shoot!' but since he was steady in his retreat I felt I could sweeten it up by closing a little more. I said 'Wait, wait' and we had closed to about 2nm when I squeezed the two Sparrows off at about a 5sec interval, calling 'Fox 1' as I fired. The rocket motor was so bright at night that it virtually blinded me for a second. The second Sparrow launched just as the first warhead detonated directly in front of us. There was a large fireball and the second missile impacted in the same spot. I came right slightly to avoid any debris. The target on our radar appeared to stop in mid-air and within a second or two the radar broke lock. The MiG-21J pilot did not survive: if he ejected after the first missile the second must have done him in. We could not see any debris because of the darkness.

"Now it is amusing, but at the time I was extremely upset to hear the controller calling range and bearing to the bogey when we reported 'lost contact' after we had splashed him. Due probably to the fairly long distances between us and the controlling ship (100nm or so) and our fairly low altitude, he had not heard our 'Fox 1' calls, so he did not know we had fired at the bogey. Bruce searched for bogeys with our radar but we saw nothing. This went on for about a minute or so until I realised that the Automatic Tracking Feature of the controller's radar had probably continued plotting the bogey's symbol on the last known course and speed when radar contact was lost. Once we realised we were tracking a 'phantom bogey' we turned east and coasted out, knowing we were in a fairly hot area – lots of surface-to-air missile and AAA sites – just south-west of Thanh Hoa. We coasted out with lots of gas (about 9,000lb) and returned to the ship for the ever-exciting night carrier landing. The kill was confirmed about three days later. The F-4J we were flying, BuNo 157299 and VF-103 side number 206, was lost in 1976 as a result of a fire while airborne in the Roosevelt Roads operating area. It was assigned to the 'Red Rippers' (VF-11) at that time, and the skipper, Cdr Rod Karber, was flying it with RIO Lt Rick Merker. They both ejected safely."

Tucker believes that the American pilots had distinct advantages over the North Vietnamese, though they were of course partially offset by the distance from home and opposition from mutually supporting SAMs and AAA. The proven capability of the Sparrow missile in the forward hemisphere was a big psychological plus, even if the probability of kill was low. The US aircrews had a great deal

**Below: The curse of the F-4: long streamers of black smoke came from the otherwise perfect J79s at any point between idle and afterburner.**

**Bottom: RF-4Bs like this VMFP-3 bird were to become the oldest Phantoms in front-line Navy/Marine Corps service.** *(Don Linn)*

more training and flew 25-40hr a month consistently, much more than their adversaries. Still, the early years of the war revealed that the crews were not adequately trained in the hard facts of air combat manoeuvring, and it was not until pilots began to emerge from the Navy Fighter Weapons School ("Top Gun") at NAS Miramar, California, that an improvement could be noted.

While the MiGCAP missions were the most exciting and sought after, BarCAP was the most tedious. The Barrier CAP station was manned continuously by a section of fighters throughout the Vietnamese conflict. The station was a 20nm racetrack course parallel to the coast and located about 20nm south-east of Haiphong. The purpose of BarCAP was to defend against surprise MiG attacks on US ships, or on special-duties or reconnaissance aircraft.

The BarCAP normally departed 15min before the scheduled launch, since the station was some 125nm from the carrier. The F-4 section would launch, take on about 2,000lb of fuel from the tanker, and proceed north to the station. Relief on station was required. The section would hold at 270-300kt, flying a combat spread. At night, or in instrument flight conditions, a 2nm radar trail formation would be flown at 15,000-18,000ft. There would rarely be anything going on, but occasionally an unidentified radar contact would pop up and the BarCAP would investigate. Several missiles were fired by BarCAP aircraft, at high-speed surface contacts, helicopters, cargo aircraft and even at a MiG or two.

About 45-60min after the section had arrived on station the BarCAP tanker would show up and give 3,000-5,000lb of fuel to each F-4. This meant that they could maintain a 2,000lb "combat package" of fuel throughout the cycle. When the relief BarCAP arrived the departing F-4s would use their 2,000lb of extra fuel in a brief one-on-one combat training engagement and then hustle back to the carrier to recover.

Below: An F-4B of VMFA-115 in the revetment area at Da Nang in 1971. Da Nang was in perpetual danger of rocket or mortar attacks, and sometimes satchel charges were planted. Revetments were essential even though space was critically short on the field. (Jim Gatewood)

Bottom: The "Black Aces" of VF-41 have one of the most famous insignia among the heraldry-rich squadrons of the Navy. This F-4B is from USS Independence. (Fred C. Dickey Jr)

Left: Aboard USS *Constellation* with VF-143, which has what is probably the worst nickname of any squadron in any air force. VF-143 somehow took on itself the name "Puking Dogs". It must have been done in a spirit of fun, but it loses something in the translation. *(Meehan)*

Below: VMFA-232 F-4J. At one time the tail lettering was "pierced" by a diabolical trident. *(Fred C. Dickey Jr)*

Bottom: Phantom from VMFA-122 emblazoned with an outsize broadsword. Note the well-used appearance of the radome. *(Fred C. Dickey Jr)*

The Navy's use of the F-4 throughout the war was far more flexible than that of the Air Force. The Air Force tended to follow rigid standard operating procedures which required entrance over certain points, procession to other points, and then flight on to the target, a process which lent itself to easy countermeasures by the North Vietnamese. The security at 7th Air Force in Saigon was considered by many to be non-existent; the North Vietnamese defences were always alert for the attack. Each Navy air wing commander had a far greater degree of flexibility in planning his attack. Offsetting this was the fact that the North Vietnamese had excellent radar coverage of all the Navy attacking routes, and it was virtually impossible to surprise them.

Yet just as the USAF changed its F-4 tactics as the war dragged on, so too did the Navy. The following three accounts of air-to-air combat illustrate this evolution. The first relates the first Marine victory, on December 17, 1967; the second is the tale of an unsuccessful but hazardous encounter by Gene Tucker; the third is an account by Navy RIO ace Willy Driscoll of the epic May 10, 1972, engagement which created the first Vietnam War aces and resulted in seven MiG-17s and one MiG-21 being shot down.

(May 1972 was a good month for F-4s in both the USAF and the Navy: the USAF shot down a total of 11 MiGs, launched the victory series of Chuck DeBellevue and Steve Ritchie, and added the second to Jeff Feinstein's sequence; the Navy shot down a total of 16 MiGs during the month.)

Capt Doyle D. Baker USMC was stationed with the Air Force's 13th TFS on an exchange assignment at Udorn, Thailand. On December 17, 1967, he was flying fighter cover for a 40-aircraft strike. Baker spotted a MiG-17 coming in on the opposite heading to the strike flight; he jettisoned his external tanks and rolled in on the MiG from about 10,000ft. The MiG saw the Phantom section attacking and entered a tight turn. Baker selected his centreline 20mm gun pod and opened fire with short bursts; the MiG kept flying lower and lower, and after four passes Baker had fired off all his ammunition. He then selected the Hughes AIM-4 Falcon heat-seeking missile and fired it right up the MiG's tailpipe. The MiG pilot ejected and enemy gunners began firing. Baker and his GIB, Capt Jack Ryan, pulled up and out. (Capt Baker later transferred his commission from the USMC to the USAF.)

Capt Tucker's story describes a more complex if less successful engagement: "On June 22, 1968, I was flying a MiGCAP mission with my radar intercept officer, Lt JG Cosmo Salibello, as wingman to then Capt Charlie Wilson, a USAF pilot on exchange duty with our squadron. We were attached to Fighter Squadron 33, commanded by Cdr Bill Knutson. We were part of Carrier Air Wing 6, led by Cdr L. Wayne Smith and flying from the deck of USS *America* (CV-66), commanded by Capt D. D. Engen.

"It was mid-afternoon, as I remember, and we were holding feet wet off the coast of North Vietnam just east of Vinh in the vicinity of Hon Mat Island. We were holding at 10-15,000ft. Our configuration was standard for F-4J MiGCAP: centreline tank, two AIM-7Es on stations 3 and 7, and two AIM-9Bs on one of the wing stations, either 2 or 8. Interestingly, it was the first combat deployment (also the first deployment ever) for the F-4J with its AN/AWG-10 pulse-Doppler radar. At that time the war against North Vietnam was restricted to the area south of latitude 19°N.

"A section of MiG-17s was detected heading south and the US controllers passed the codeword over guard frequency for all friendly aircraft to vector feet wet so there would not be any confusion causing friendly-versus-friendly engagements. Shortly thereafter our section was told to vector for Blue Bandits (MiG-17s) 35-40nm north-west of us, headed south-east. We headed due west for the first minute or so to get on their flightpath before turning north-west.

"As we coasted in I could see an A-7A coming directly at us, headed towards the Gulf of Tonkin. The pilot was Lt George Webb, later to be CO of VA-81, who, upon seeing us headed right at him, was concerned we might have him confused with the MiGs. In those days the J79 engine was a bad 'smoker,' which made F-4Js at military power easy to

**Below: A Phantom from VF-302 takes on fuel from a Douglas KA-3B. This modified Skywarrior could carry up to 5,026gal of fuel, of which more than 3,000gal could be offloaded in mid-air.**

sight under most conditions. He switched his radio to guard and transmitted: 'Don't shoot, F-4s, don't shoot!' Good headwork on his part, though unnecessary, as we saw the MiGs on our pulse-Doppler radar and Cos, a superb radar operator, had them locked up bearing 310° at 25nm as we crossed the beach line at about 3,500ft. We turned to 310° and proceeded directly at them, a track which took us right up the Son Ca River valley. We were cleared to fire at that time and were all armed up. We had elected not to jettison our centreline tanks since we were above jettison speed from the time we received our vector.

"Capt Wilson had the lead until we made the turn, but his RIO did not have radar contact, so we took the tactical lead at that time. We were in a combat spread formation with Charlie about 1¼ miles on my right beam and stepped up 1,500-2,500ft.

"We closed 'straight down the pipe'. The weather was very clear; visibility was super and I got a tally-ho at about 10nm (honest!) and could tell they were a section of MiG-17s at 7nm. I was amazed to see that not only were they continuing straight at me and into the envelope of my trusty forward-firing Sparrows, but that they were flying so close together that they actually looked like they had wing overlap – like the Thunderbirds or Blue Angels. The wingman had to have nearly 100 per cent of his attention devoted to his lead he was so close. I recall thinking they were dumb to come straight at me since I had a forward kill capability and they didn't (except for a minimal head-on gun-shot opportunity), and that if they remained in that formation I could get them both with one missile, not that I intended to try only one. We had selected minimum afterburner at about 12nm, and were doing about 550kt.

"We continued to close and were in range at about 6-7nm, but had decided to wait until approaching mid-range. At about 4½nm we were at optimum range. They were about 2° left of my nose with a steady bearing and fire-control solution as I squeezed the trigger to fire a Sparrow with all systems indicating up. The 1.4sec between trigger squeeze and a Sparrow coming off the fuselage station is the longest

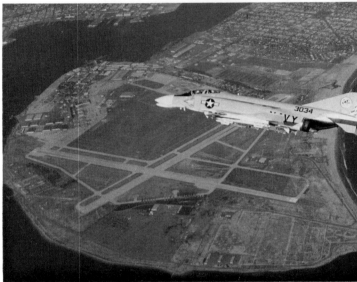

1.4sec in the world. I squeezed the trigger to fire the second Sparrow at about the time I realised there was some problem with the first missile. It didn't take me long to realise that neither missile was working properly, although at the time I didn't know why. At our altitude and airspeed – low and fast – there was a moderate airframe vibration, and I did not feel the missiles leave the airplane, but in fact they had. Unfortunately, I was to find out later they had been improperly loaded. A cotter pin had been inadvertently left off where the upper motor-fire connector lanyard was attached to the fuselage station ejector foot. As a result, the motor-fire voltage was never passed to the missiles, and they were in effect jettisoned.

"The situation had changed slightly. Instead of the MiGs dodging my Sparrows, they now had me in their sights if the pilots were trained that well. I continued to close them head-on, but now watching closely to see if I could see any signs of them shooting guns at me. To this day I don't know

**Above left:** Coats of many colours worn by aircraft from VF-201, VF-202, VF-301 and VF-302.

**Left:** The Naval Air Rework Facility at North Island, San Diego, California, was responsible for creating this "new" F-4J. *(US Navy)*

**Above:** F-4J from VMFA-333, photographed on July 10, 1970, at Randolph AFB, Texas. "Double Nuts" (two zeroes) indicate that the aircraft is flown by a Commander, Air Wing.

if you can tell from virtually head-on with 800 or so knots of closure, but I was sure looking. We passed within less than 50ft of the left side of each other, looking 'eyeball to eyeball'. And as Randy Cunningham likes to say, 'I could see their little Gomer hats, and their little Gomer goggles.'

"I immediately started a nose-high port reversal, selecting full afterburner. In the meantime, my wingman had turned into them, and as I passed the MiGs he had 90° on them and they were on his nose. Unfortunately, although his system was working, he still did not have a radar contact. He called for his RIO to select boresight (slave the antenna to the nose/gunsight) and he put the MiGs in the gunsight. He squeezed the trigger and his two Sparrows fired, but they went unguided as he had fired before his RIO actually got the boresight mode selected.

"At the time I couldn't figure out why he didn't roll in on their tail, but I now believe those MiGs turned into him and departed to the north as he overshot due to the large turning radius of the F-4. I had lost sight of the MiGs as I started my left reversal, but after completing about 90° of turn I visually picked up Charlie in a left turn at about my 8.30, 4-5nm, but he had a MiG-17 at his 7 o'clock, about a mile. I told him about the MiG and told him to keep coming left, planning to continue my reversal and shoot the MiG from Charlie's tail. As I continued my turn, wondering how the MiG got on his tail so fast, I realised that the MiG on Charlie's tail could not have been one of the MiGs I had passed – that there were more around. Feeling uncomfortable, I overbanked to the left, checking inside my turn. I saw nothing and reversed my bank to the right, checking my belly-side. Sure enough, there was a MiG-17 about 1¼ miles at my 4.30 low. He was about 30° nose-high, pointed straight at me. I estimated his speed at 350kt and decelerating, and figured he wasn't carrying Atolls: I didn't see any on him (but at 1¼ miles, who can?) and many MiG-17s were loaded guns-only in those days. I decided that (1) my MiG was not a big threat to me – out of range and decelerating while I was accelerating through about 550kt at the time; (2) if I turned into my MiG I would have to go several turns at best to get a shot; and anyway (3) I was needed more by Charlie because of his MiG. I made my decision to go for Charlie's MiG.

"I reversed left, picked up Charlie and his MiG at about my 9 o'clock, about 4nm, and continued pulling hard down into them. They were at 3-4,000ft and I had peaked out at 8-10,000ft altitude. They had turned through another 60° or so of turn and I was trying to figure out how to prevent a big overshoot while trying to get to the 6 o'clock of Charlie's MiG. Right about then, a white trail of missile motor fire smoke passed high over my canopy from about 7.30 to 1.30. I broke hard left and up into the threat and saw a silver MiG-21 up at around 20,000ft going the speed of heat – I estimated 1.3 or faster – in a left level to slightly nose-up turn. I was down around 5,000ft at the time – the Atoll had not even come close to me. The MiG continued his turn and disappeared towards the north before I could get him in my sights, although he was probably well out of the envelope with his altitude and speed advantage.

"Returning my attention to Charlie, I looked down and found him at my 11 o'clock, 2nm, as I was headed about east, coming right at me. There was no longer any MiG behind him. Charlie and I passed each other, ensuring our sixes were clear. We looked for the MiGs briefly while trying to get together in a tactical formation. We saw no MiGs but unfortunately lost sight of each other again. I said something astute like "Let's get the hell out of here!" and he agreed. We both turned east – I unloaded and accelerated – and headed for the Gulf, maintaining radio contact and checking our

Left: Marine training squadron Phantoms. In Air Force argot, "SH" would be the perfect tail insignia. (USAF slang for something really good was "Sierra Hotel".) *(MDC)*

Centre left: First flight of the F-4J. Note the stabiliser leading-edge slots, made necessary by the new wing slats. No 153072 was actually the third aircraft in the J series.

Botton left: A laden Phantom from VMFA-314. Squadron commander Norm "Animal" Courley has his name inscribed in modest letters on the spine of the aircraft. Skyhawk from sister unit in background.

own sixes. I was around 5,000ft when I unloaded, and ended up on the deck – at about 100-200ft going about Mach 1.2 (over 700kt) – and really smoked out of there. My route, I realised at the last instant, took me almost directly over the citadel of Vinh. If there were any unbroken windows or china in Vinh up to that time, there could not have been after I passed. I'm sure I had one hell of a shockwave. I knew I was fast – the canopy was too hot to touch – but I didn't realise how low till I crossed the 'gooseneck' of the Son Ca River east-south-east of Vinh and passed beside a small sampan/rowboat with a Vietnamese fisherman standing in it. I felt like I was almost level with him. In truth, I must have been at 50-100ft because I was looking up at Hon Nieu Island as I coasted out, and as I remember it is only about 200ft high. After we were well at sea I slowed and climbed, Charlie and I joined up, and we headed for the carrier.

"In retrospect, though the controllers only saw one section of MiGs, in fact there was a second section of MiG-17s 3-5nm in trail at a low altitude, and a MiG-21 sitting high over the fight. We should have seen the second section sooner, but got so engrossed in the lead section that we never considered the possibility of a set-up. However, they were developing that tactic and used it a lot for the next couple of years. After we figured that out, we [Navy fighters] decided that we would try to have the wingman keep looking for a second section with his radar, and upon closing the first section fly through to the possible second section rather than engage the first section and end up turning in front of an unknown second section. In actual fact, when most fighter pilots saw any MiG, they ended up turning with that MiG. 'A bird in the hand' was the philosophy.

"The second section of MiG-17s were probably too far behind: if they had missiles they never got into the envelope, although Charlie's MiG was sure nibbling on it. If they didn't have missiles, they weren't even close to a shot. The MiG-21 pilot just flat blew it. He must have been inexperienced and scared because he surely could have had me. All he had to do was point his nose down at me as I was pitching back on the first MiG section, and close me to get in the envelope. I would probably never have seen him. I was indeed fortunate that he was a 'plumber'. Instead we all lived to fly and fight another day."

The best account of the third action, the brawling May 10, 1972, air battle, is in an official, formerly classified document entitled "An Introduction To Air Combat Manoeuvering (ACM)". The book was written by fighter pilots for fighter pilots, and combines a breezy aggressiveness with the under- and overstatement that characterise fighter instructors. Its hectic, breakneck style would never win a Pulitzer Prize but nonetheless rivets the attention of young fighter pilot trainees: "Two of the Navy strikes that afternoon [May 10, 1972] were targeted on the Hai Duong railroad bridge and Hai Duong railroad yards. One of the strike groups had a normal encounter, and two F-4B MiGCAP bagged a MiG-17. The other strike group had a flail. Some large number of MiGs – maybe as many as a dozen MiG-17s, four MiG-21s and two MiG-19s – decided to share the same airspace with the gaggle of A-7s, A-6s and F-4s. In the ensuing minutes Steve Shoemaker and Keith Crenshaw killed one MiG-17, Matt Connelly and Tom Blonski killed two and Randy (Duke) Cunningham and Willy Driscoll killed three (with three Sidewinders). Our side didn't lose any airplanes to MiGs. The kills made the latter crew the first aces of the Vietnam War. Randy has told his story in print at least twice, so this is Willy's narrative. Willy deliberately omitted any comments about the tactics employed and has attempted to clearly describe what he saw, heard and felt (but minus the emotion).

"On May 10 Duke and Willy and their wingman (Brian Grant and Jerry Sullivan) were flying as flak suppressors. Their F-4s carried four Rockeye, four AIM-9, two AIM-7 and a centreline.

"On the way in there was only sporadic AA; no SAM firings or barrage AA were observed. Willy commented to Duke that this lack of activity was unusual. However, neither of them thought that this might mean that they'd see MiGs. There were no MiG warnings from Red Crown.

"The strike leader tallied the target and the attack began, with most of the bombers making west-to-east runs. Visibility was exceptionally good as Duke and Willy made an easy port orbit on the outside of the strike group. During this time their visual scans were directed exclusively at the ground, looking for muzzle flashes. Since they saw little flak activity, they set up for a bombing run on a warehouse near the railroad yards. Duke commented to Willy that everyone was rolling in from the same direction, so they would go north-to-south to avoid being predictable. At this point both of them were still concentrating their scan on the ground. Red Crown still had not issued any MiG warnings. As they rolled in they saw that their warehouse had already been severely hit, so Duke picked up a smaller structure to the west. Now the flak started up, and Duke released steep, fast and high. As they pulled off to the south, Willy saw an F-4 at his left nine high on the east-to-west run surrounded by heavy flak. Seconds later it became an orange-black fireball. He then saw one crewman eject. Duke continued a port nose-up turn from south to south-east to look back at the target.

"While Willy was looking down and back to see their bombing accuracy, a flash in his peripheral vision revealed a MiG-17 at left seven low, one and one-half miles and closing. He called this to Duke, who retransmitted it over UHF and told Brian Grant he was going to drag the MiG east for him. Duke and Willy were now at 330 KIAS after their jinking pullout. While Duke was transmitting, Willy sighted four more MiG-17s on the left, very low. The MiGs were moving from a tight left echelon into two elements of two aircraft. Willy thought that the first two MiGs were going to attempt a gun pass as they dug further to the inside of Duke's turn and closed. The other two MiG-17s continued wings-level in tight formation and disappeared through Willy's right six. Duke had the airplane in an easy port, wing-down, nose-down acceleration. Willy now told him that the first MiG-17 was inside 2,000ft and still closing. Duke then told Brian Grant to shoot that MiG. Grant replied that he had three MiGs on him and couldn't, whereupon Duke broke hard port obliquely down into the threatening MiG. Willy saw that the MiG was making no attempt to continue the attack and told Duke that it would undershoot badly. It did, and Willy lost sight of it as it flew under the belly of the F-4. He then switched to the other two MiGs, which were in a hard port turn trying to maintain their position of advantage. They were about 2,000ft away, but their fuselages were not aligned with that of the F-4. Duke immediately regained sight of the MiG which had badly overshot. It was at right 1 o'clock, slightly high in an easy port turn. Duke rolled starboard and then reversed port in a 5-6g turn, which put him at the MiG's right six, slightly low. Duke squeezed off the first Sidewinder and it flew up the MiG's tailpipe. The MiG was almost straight and level in afterburner. Willy saw the orange-black fireball in his peripheral vision, but he was more concerned with what was going on between his left nine and left six.

"Duke's reversal to port let Willy regain sight of the other two MiG-17s. They were at left seven level trying to close for a gun pass. Willy called this to Duke, who pulled 5g up and port. During the vertical disengagement Willy tallied four more MiG-17s on the left, about 6,000ft below at tree-top level in a port turn. Duke rolled starboard for a belly check, and Willy saw at least another four MiG-17s. They too were low in a level port turn. Since the bombing run Willy had been looking aft between right three and left nine, and didn't have a clear picture of what was going on in front of the F-4. Now, as they started nose-down port, Duke called that he had tally on several MiG-17s at his right 1 o'clock low. The lead MiG was approaching a guns-tracking position on the XO. That F-4 and all the MiGs were in level port turns, drifting towards Duke's left ten, slightly high. Duke pulled the F-4 into a hard nose-up port turn, and Willy lost sight of the MiG-17s at his left seven low. These MiGs had closed inside 2,000ft on what appeared to be a possible gun-tracking solution. Willy called this to Duke, who rolled port, allowing Willy to reacquire these MiGs. One was at left six low, and the other two were at left seven in a very tight fighting wing. All three were inside 3,000ft, but none appeared to be closing. Willy warned Duke about a possible gun attack just as the two MiGs at left seven began shooting. The 37mm nose gun in each was sending out distinctly visible white rhythmic muzzle flashes. Then both MiG-17s simultaneously began firing their 23mm guns. The 23mm flashes weren't as large, but their firing rate was nearly twice that of the 37mm. Willy told Duke that they were shooting but that Duke's hard port turn was giving the MiGs sufficient tracking problems. Neither MiG attempted to close inside 2,000ft. They had difficulty stabilising their aircraft and they were very close together. Willy thought that their wings were close to overlapping. They were bobbling up and down like Training Command students trying to fly formation. Because of all this, Willy didn't call for any additional guns defence manoeuvres.

"Meanwhile, Duke was attempting to close and shoot the

**Left:** VF-302 Phantoms going straight up.

**Top:** Again, the markings leave no doubt as to who is flying what aircraft. The commander of Carrier Air Wing 30 is in the lead of this VF-302 two-ship.

**Above:** Two RF-4Bs from VMFP-3 streak in low over the desert. These aircraft have done sterling service with the Corps. *(Harry Gann)*

**Right:** A VF-154 Phantom from the USS *Ranger* offloads its bombs. When the bombs are gone, the missiles can be used for air combat. *(MDC)*

MiG-17s pursuing the XO's F-4. He called for a port break three times, but the XO kept turning port without breaking. Duke called that an F-4 was going to get gunned down and tightened his turn. This compounded the tracking problems of the two MiG-17s at left seven, and Willy didn't feel threatened by them. Then Duke relaxed the turn. The two MiG-17s overshot mildly, but now the third MiG (initially at left six) rolled into a 45° bank and pulled lead while closing with his fuselage aligned with the F-4. Willy called to Duke to tighten the turn as this MiG closed inside 2,000ft, well to the inside of their turn. The MiG began firing his 37mm gun. Passing through 1,200ft and still closing, he began firing his 23mm guns. Willy could now distinctly see the metal rivet lines on the MiG, the boundaries of its grey-purple camouflage, and a bright red star on each outer wing panel. Willy told Duke that a MiG-17 was in a valid guns-tracking position on the port side with its guns blazing. As he did, he heard the distinct exhaust whish of a Sidewinder launch. He started to scream this to Duke when Duke snap-rolled the F-4 starboard. Willy lost sight of this attacking MiG-17 when it was inside 800ft, still closing and still shooting. As Duke unloaded and extended to the south, Willy saw and called several MiG-17s to starboard, none a threat.

"Duke reversed port and Willy reacquired the attacking MiG-17. It was also in a roll to port, but its bank angle was lagging by about 120°. As the MiG approached the F-4's bank angle, Duke again snap-rolled the F-4 to starboard and unloaded to accelerate. When Duke rolled back to port, Willy regained sight of the MiG-17, now trailing by 4,000ft as a result of the defensive roll-aways. Somebody called Duke and told him not to re-enter the fight because he had numerous MiGs on both right and left. Duke rolled the F-4 into a 90° port bank and continued to accelerate while Willy alternated watching the MiG-17 whom they'd just outrolled and looking for other MiGs on the port side. That MiG-17 was the only enemy fighter visible on the port side. Willy had difficulty maintaining sight of the head-on MiG-17 at 1 mile. Before starting a port climbing turn towards the east, Duke rolled the airplane starboard so Willy could check that rear quadrant. It was clear. Willy now thought the fight was over.

"Duke began an easy port climbing turn towards the coast. Almost immediately he told Willy that he had

Above: Radar intercept officer LtJG William "Willy" Driscoll (left) and Lt Randall Cunningham pictured after shooting down their second MiG-17. Two days later they downed three more to become the first US aces of the Vietnam War. *(US Navy)*

Right: Cunningham and Driscoll arrive back aboard the USS *Constellation* after they had been fished out of the Gulf of Tonkin by the destroyer USS *Samuel Gompers*. They had just destroyed three MiGs before themselves being shot down by an SA-2 missile. *(US Navy)*

something on the nose that looked like a MiG-17. Then he said he was going to scare the MiG pilot with a head-on pass. Willy kept checking left and right six. Then he felt the F-4 reverse into a hard starboard turn. (Duke later told him that the MiG had started shooting head-on so he reversed starboard to avoid the MiG's bullets.) Duke next pitched port hard, nose-up. At his left eight level Willy clearly saw this MiG-17, also in a port nose-up turn. The MiG had a faded green-and-brown camouflage scheme. It was 180° off the F-4's tail, well inside a half-mile. The MiG pilot used an extremely nose-high port rolling manoeuvre to establish and maintain a position well inside the F-4's turn. Duke told Willy where the MiG was and what he was going to do while Willy checked left and right six and told Duke they were clear. They talked back and forth like this through two additional vertical rolling manoeuvres.

"Each time the MiG appeared in Willy's scan, it was either in a nose-high or a nose-low rolling manoeuvre, maintaining its position on the inside of the F-4's turn with its nose on the F-4. Willy thought several times that this MiG pilot flew a competent aggressive airplane, unlike his contemporaries.

He had no reservations about using the vertical, and he rolled skilfully to maintain his advantage while in the vertical. He was also able to keep his fuselage nearly always aligned with that of the F-4. On the last port vertical manoeuvre with the MiG-17 at left seven slightly high in a port nose-down turn, the MiG pilot suddenly relaxed g and rolled wings-level. As Willy was calling this, Duke rolled the F-4 nose-down onto the MiG, somewhat like a low yo-yo to port. The MiG pilot flew straight and level north towards Hanoi. Willy told Duke that it looked like a bugout as the MiG moved forward of left nine low. The F-4 was in a 90° left bank when Willy heard the exhaust whish of a Sidewinder. He didn't see what happened because he was checking left and right six. As he did this he felt the F-4 continue port and nose-down almost to tree-top level. Duke was following the MiG down to make sure that Sidewinder was really a lethal hit. Seeing that it was, Duke rolled the airplane hard starboard 90° and started climbing. Willy asked him for fuel and he reported 5,500lb.

"At this point Willy started hearing UHF transmissions for the first time since the start of the fight. The rest of the

strike group was reporting feet wet and switching to the nearest tanker frequency. Duke and Willy were 10 miles from the nearest strike aircraft and 40 miles inland. Willy also noticed that Red Crown was calling numerous MiGs near Hanoi. Willy had been checking both left and right six, but the MiG warnings prompted him to concentrate on left six. Neither Duke nor Willy noticed any SAM indications on their warning gear, and they didn't expect any because their AIs at this time had briefed them that the North Vietnamese didn't mix the two. They both felt relatively relaxed.

"As Willy scanned from left six to nine for enemy aircraft, he saw an exceptionally bright white flash on the right side of the F-4 in his peripheral vision. It was like turning on the bathroom lights in the middle of the night after awakening from a sound sleep. Almost immediately there was the sound of small metal pieces hitting the side of the F-4 – like someone had thrown a handful of BBs against the side of the airplane. Willy quickly swung around to look right and aft. He saw the unmistakably bright orange smoke of an SA-2 warhead detonation. It looked like it was inside 500ft. Willy asked Duke about his ECM indications as soon as he saw the orange smoke. Neither Willy nor Duke had noticed any ECM indications for the last 30sec, even though the gear had worked perfectly earlier in the mission. Duke said that a SAM had just gone off close on the starboard side but that the engine gauges were OK and the airplane didn't appear to have any problems. Willy told Duke that he'd seen the detonation and that although it looked and sounded lethal, he thought it was outside max lethal range. Then he started clearing six again. Suddenly the F-4 yawed hard port and down. Willy thought Duke had sighted another MiG and he started rotating his body from aft to forward, scanning intently for MiGs.

"Duke answered that he had utility hydraulic problems. The F-4 continued to yaw excessively to port, and Duke said they were rapidly losing the utility hydraulic system. Willy immediately concluded they'd need a carefully set-up arrested landing, probably not on a carrier. Therefore he pulled out his bingo fuel card and looked up the altitude, power setting and descent range for a bingo profile to Da Nang. At this point he was only mildly concerned, and they continued to climb through 17,000ft, about 20 miles from the coast. Then the F-4 yawed port hard again, and this time tucked into a 90° left bank. Duke said they were losing flight

**Top left: Cunningham and Driscoll in the ready room after their ace-making MiG kills.** *(US Navy)*

**Centre left: Lt Michael J. Connelly, an F-4J pilot from VF-96, explains his tactics in downing two MiGs as his radar intercept officer, Lt Thomas J. Blonski, looks on. Date was May 5, 1972.** *(US Navy)*

**Left: VF-96 executive officer Cdr D.O. Timm congratulates Connelly after his victories. Lt Blonski looks on at left.** *(US Navy)*

**Right: Lt Steve Shoemaker (foreground) and LtJG Keith Crenshaw after their MiG victory on May 10, 1972.**

control hydraulic pressure now. Willy then realised that there was nothing Duke could do with both systems failing. They were going to have to eject, and the only question was whether it would happen feet dry or over the Gulf. While adjusting to this thought, Willy felt the F-4 yawing hard and continuously to port. The yawing gyrations didn't exceed a 45° bank, but did decrease the F-4's altitude. Climbing was impossible and level flight became increasingly more difficult as the yawing increased in frequency and intensity. Willy looked over his right shoulder towards Nam Dinh several times, but didn't see any additional SAMs or fighters. As the yaw intensity built up, he began to spend more time monitoring the gauges and less time clearing six. Willy also double checked the command ejection handle to be sure it was set for him to command the ejection. He told Duke it was properly set. Duke told him to stay with the aircraft until it reached the coast. Willy had already made up his mind to do this since his fear of being a POW (especially after five MiG kills) overrode his fear of death.

"Duke said he thought they'd make it to the coast, but it looked to Willy like the coast was 10 miles off, and the F-4 was now almost uncontrollable. As Willy looked out he could easily see intense fire surrounding the fuselage and inner wing on both the right and left sides. The violent yawing continued for about a minute as speed and altitude decreased. Just as the F-4 was directly over the coast, at 13,000ft and 200kt, there was a hard nose-down port yaw. Willy thought the F-4 had entered a spin. Airspeed bounced between 50 and 80kt, but there was no negative or zero g. The sky, sea and land revolved around a 360° sphere. Willy looked down at the secondary handle, put his hands on it and got into the right ejection position just as Duke told him to eject. His first sensations after ejection were of the tremendous calm and peaceful nature of this new environment. He next looked up at his fully blossomed orange-and-white parachute. Duke was also in a good chute about half a mile away. The F-4 was burning below, and as he watched it Willy suddenly saw the F-4 burst into a blackish-orange fireball.

"They landed about a mile offshore and spent about

20mins there until rescued by Marine helicopters from an LPH. Both Duke and Willy received the Navy Cross for this mission.

"As we said, Willy deliberately removed the emotion, as is natural for a skilled ACM instructor. Nevertheless, it is hard to miss the terror, relief and elation buried in this clinical account."

Cunningham and Driscoll were flying an F-4J from VF-96, operating off the *Constellation*, when they became the first aces of the war.

The Navy's air war in Vietnam has, like the veterans who returned from it, received shabby treatment from history. It was a titanic struggle in terms of logistics, skill, bravery, persistence and achievement. More than a dozen carriers – *Enterprise, Intrepid, Bon Homme Richard, Midway, Ranger, Saratoga, Coral Sea, Hancock, Oriskany, Franklin D. Roosevelt, Forrestal, Independence, Kitty Hawk, Constellation* and *America* – had fought a valiant battle for almost a decade, hampered by the same rules of engagement which had reduced the effectiveness of the USAF. Pilots had been required to fly as many as 25 to 30 combat missions per month, month after month, deployment after deployment. At war's end some had logged as many as 500 combat missions against some of the most sophisticated air defence systems the world has ever known.

Through it all, the Phantom demonstrated a versatility and growth capability never before seen in a combat aircraft. Oddly enough, despite the outstanding success of the Vought F-8 Crusader's gun weaponry, the Navy did not procure an internal gun-equipped F-4 as the Air Force did. Still, it seems appropriate that the last MiG to be shot down in the Vietnam War was splashed by a Phantom, just as the first had been. Both victorious Phantoms flew from the *Midway*, the first an F-4B from VF-21, the last an F-4B from VF-161. Lts Victor Kovaleski and Jim Wise caught a MiG-17 over the northern part of the Gulf of Tonkin on January 12, 1973, and dispatched it with two Sidewinders. By a curious twist of fate, Kovaleski, this time with Ensign Dennis Plautz as his RIO, was the last man to be shot down over North Vietnam, being gunned down by anti-aircraft fire on January 14. He and Plautz were rescued by helicopter.

There is one final point to be made about F-4 operations with the Navy. No matter how boring or how exciting the mission had been, at the end of the flight you had to get back aboard the carrier, day or night, good weather or bad. The nearest divert field was Da Nang, some 200 nautical miles from Yankee Station, too great a distance for an F-4 which had burned its fuel down to the maximum landing weight unless there was a tanker airborne.

Each aircraft has a maximum gross weight at which it can "trap" (engage a carrier arrester-gear wire). That weight is made up of the aircraft's basic weight, plus the weight of the stores carried and the fuel aboard. In the period before 1970, F-4s with a normal missile and bomb load frequently had to recover with less than 3,500lb of fuel, and sometimes, depending upon the load, with as little as 2,200lb. A typical night pattern took 800lb, so there wasn't a lot of leeway. In the 1970s the maximum trap fuel weight was raised to 5,800lb, although there was rarely this much remaining after a mission.

When the mission was over, after a massive expenditure of nervous energy in the course of a trip through a flak-ridden target, there still remained the task of hurling a 40,000lb F-4 flying at 145kt at the heaving deck of a carrier, at night, in bad weather. All of the pilot's will and energy had to be summoned up; he had to concentrate totally on the task, to place himself in the hands of the Landing Signal Officer and his instruments, and to accept the challenge. And imagine the thoughts of the equally tired RIO in the back seat, unable to see and totally dependent on the skills of a pilot he knew to be approaching the end of his resources. It called for a new dimension in courage.

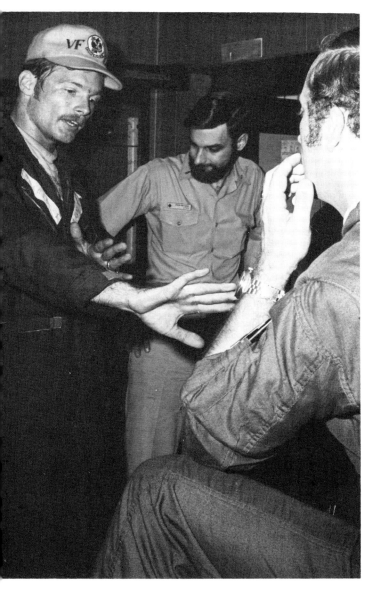

**Far left: One more time, Cunningham goes over his three MiG kills.** *(US Navy)*

**Left: Pilot Lt Curtis R. Dose (left) explains how he manoeuvred his VF-92 Phantom behind a MiG before shooting it down.** *(US Navy)*

105

# 7. Fighting the Phantom

Ever since the days when Boelcke and Immelmann invented the deadly *Kurvenkampf* ("turning fight"), through the slaughter of the First World War, the bitter battles of the Second World War and the fierce contests over Korea's Yalu River, the winning fighter was most often the one able to turn the tightest, keeping its nose swinging across and inside the track of its opponent until the guns were brought to bear. Loss of speed meant loss of battle, and the vertical dimension was only resorted to as a means for hit-and-run attacks, with a plunging dive towards a more nimble opponent and a zoom climb away.

The Phantom II introduced new concepts of manoeuvrability in the vertical plane. The Phantom and some of its contemporaries offered, for the first time, enough surplus power to convert the combat arena from a horizontal plane to an egg-shaped field of manoeuvre in which the adversaries could adopt a whole new range of relative positions. The fighter became not just a man with a gun, but an integrated system whose turn, climb, acceleration and weapon characteristics had now to be analysed to get the most out of them.

With what came to be called "air combat manoeuvring" (ACM) there came tons of theory, miles of graphs, hundreds of ribbon diagrams, and the need for F-4 crews in particular to have far more mathematical understanding and management ability than any aircraft crews in history. Despite the images and the slogans, despite their gruff insistence that they are basically simple killers who love to fight and whose only message to the enemy is "Don't fly in my sky," fighter pilots have had to become flying scientists. Their body of knowledge has to be so ingrained that they are able to react correctly and instantaneously to any one of an infinity of permutations of aircraft state (weight, altitude, speed and configuration), enemy aircraft state, munitions, missile-firing parameters, distance from home, tanker location and so on. In essence, the modern fighter, particularly the F-4, demands a mathematical analysis before, during and after the acceptance of combat. Obviously, the data needed for

**Below: Air-to-air tanking was the key to many of the USAF's Vietnam operations. The F-4D on the tanker is fitted with the Loran-D precision navigation system, the LSI ARN-92 nav-attack system, and the Westinghouse Pave Spike laser pod. F-4s thus equipped were known as "Pave Phantom" aircraft. The F-4E shown had been delivered to Seymour Johnson AFB, South Carolina, and then moved to Ubon RTAFB.** *(MDC)*

Above: Fast FAC Phantom over Route Package I, north of the DMZ. There were three distinctive lakes in the southern part of North Vietnam that Phantom crews could use as landmarks. This one was Bat Lake; others were T-Bone and Pork Chop.
*(Dr McGregor Poll)*

such an analysis cannot be consulted during combat, so they have to become second nature to the aircrew. Moreover, the decisions based on this information have to be made by individuals undergoing high excitement, extreme stress and the discomfort of high g forces.

Later aircraft, like the F-15, F-16 and F-18, have more advanced systems which do a lot of this work automatically, reducing the load on the crew. In the F-4 the aircrew must do most of the work.

Success in air combat ultimately depends on aircraft performance, and the crew's utilisation of that performance. In the F-4 much depends on the angle of attack, the angle of the wing in relation to the airflow past it. The Phantom accelerates best at an angle of attack of 3-5°; cruises at 7-12°; turns best but suffers heavy buffet between 18° and 21°; and stalls at about 27°. In the rolling plane, the F-4's aerodynamics dictate that the pilot use aileron below 12° angle of attack, aileron and rudder combined between 12° and 16°, and rudder only above that point. As we shall see later, the introduction of aileron at high angles of attack produces adverse yaw and can cause the aircraft to depart from controlled flight.

Stability augmentation, provided by a series of electronic black boxes, helps greatly in keeping the aircraft under control at high angles of attack, and permits the pilot to manoeuvre in conditions of high buffet that would otherwise signal him to reduce his angle of attack. Buffet – structural oscillation caused by airflow breakaway – is so routine in high-g manoeuvres in the F-4 that it is accepted without comment.

Curiously, the F-4's best fighting altitude against the low-wing-loaded MiG-17 and 19 is below 15,000ft. Analysis of the opposing types reveals that at such altitudes the F-4's energy manoeuvrability and weapon system combine to give the best result. The F-4 has more excess power, better instantaneous and sustained manoeuvrability, better fuel economy (because the afterburner is used less), and more firing opportunities for the Sparrows and Sidewinders. At the same time, conditions are difficult for the heat-seeking Atoll missiles carried by the MiGs, and the SAM envelope of hazard is reduced.

Even to experienced jet pilots, whose lives are guided by checklists, the F-4 has a bewildering series of numbers to remember, including maximum and minimum airspeeds, turn rates, g loads, centre-of-gravity limits, angles of attack and manoeuvring envelopes for both aircraft and weaponry. The aircraft's handling is affected by variations in speed, g load, configuration and weight, and the pilot has to take account of this when setting up his attack.

This multitude of data has been distilled down into a set of Basic Fighter Manoeuvres (BFMs), which depend for their effectiveness as much upon physics, geometry and trigonometry as on the pilot's visual acuity, situational awareness, strength, flying ability and desire to win. They are tailored to the combat situation, which can be one-on-one with another similar aircraft, one-on-one with a dissimilar type, or any larger combination of numbers and types. Further complications arise when there is a choice of means of attack (missiles or guns) and a requirement for positive identification of the enemy in an environment where friendly aircraft may be present.

Finally, looming over everything is the awesome fuel consumption of the big J79 engines; three minutes in afterburner can account for 3,000lb, forcing the pilot to break off and run for home after a single engagement.

But no assessment of the complexities of combat in modern jet fighters in general and the Phantom in particular could be complete without hearing from someone who has both fought his own battles in the F-4 and commanded other men as they did likewise. One such man is Lt-Gen John J. Burns USAF (Retd), whose fighting career spanned three generations of air warfare, starting with piston-engined fighters in the Second World War and culminating in supersonics in the skies of South-east Asia 25 years later.

He entered the Aviation Cadet Programme in August 1942 and was commissioned as a second lieutenant in December 1943. He flew 106 combat missions while assigned to the 371st Fighter Group, operating Republic P-47s out of bases in England and France.

In November 1950 Burns went with the 27th Fighter Escort Squadron to Korea, where he flew 102 combat missions in the Republic F-84 Thunderjet. In between his various combat tours Burns worked in operational units at successively higher levels, bringing the first-hand experience of air fighting to his command positions. In July 1957 he became commander of the 522nd Tactical Fighter Squadron, the first unit to be equipped with McDonnell F-101 Voodoos. He took this squadron to Okinawa in September 1958, during the Formosa crisis.

After further command experience at squadron level with the 91st and 92nd Tactical Fighter Squadrons, he became director of operations of the 4th Tactical Fighter Wing, Seymour Johnson Air Force Base, North Carolina. His first

**Above: An F-4E with Pave Space designator pod, smart bombs and rocket launchers.** *(William Vasser)*

**Right: An ECM pod-equipped F-4D from Udorn Royal Thai Air Force Base photographed in September 1971. The Americans' ability to create effective airborne electronic countermeasures equipment was central to their success in the hostile skies of North Vietnam.** *(USAF)*

Above: An underside view of a fully loaded F-4D. This aircraft is carrying 17 500lb Mk 82 Snakeye bombs, an ECM pod, a KB-1B strike camera, two AIM-7E Sparrow missiles and a 600gal centreline drop tank. Take-off weight was 58,000lb.
(Albert Piccirillo)

experience with the Phantom came when he was assigned to Edwards Air Force Base as commander, Detachment 2, Headquarters, 831st Air Division, charged with Category II testing of the F-4 and the Northrop F-5.

After further experience at Tactical Air Command Headquarters Burns was assigned to Ubon Royal Thai Air Force Base, Thailand, in May 1967, where he served first as deputy commander for operations of the 8th Tactical Fighter Wing; in December of that year he became vice-commander. During his tour of duty he flew 132 combat missions, 80 of them over North Vietnam.

Burns held a number of important positions at Headquarters USAF from July 1970 through 1973, when he was assigned again to Tactical Air Command as commander of the Twelfth Air Force, Bergstrom Air Force Base, Texas. He assumed command of the United States Support Activities Group and Seventh Air Force (USSAG/7AF) at Nakhon Phanom Royal Thai Air Force Base in September 1974, and while there directed the military support of the evacuation of Phnom Penh and Saigon, and the recovery of the *Mayaguez* and its crew.

He subsequently became deputy commander-in-chief of US Forces in Korea, and was then appointed deputy commander of the United States Readiness Command. He retired in 1979 with over 35 years of tactical fighter operational experience and joined McDonnell Douglas as vice-president, advanced engineering.

This biography does not do justice to the hours he spent in the cockpit in combat, and the even more agonising months and years in combat commands, responsible for both the success of the mission and the lives of the men he commanded. The F-4 is a favourite of his, because it allowed him to achieve success in both areas. But as fond as he is of the Phantom, Burns is aware of its bad points as well as its good. His assessment is that of a fighter pilot who flew the type in combat, a commander who led other fighter pilots to war, and an engineer who understands not only what is desirable but also what is possible.

Gen Burns regards the F-4 as the premier air combat fighter of the time, with good speed, payload, armament and thrust-to-weight ratio. It was a tolerant aircraft, easy for pilots to fly and commanders to dispatch. This quality resulted partly from the thrust-to-weight ratio and partly from the generous wing area, a legacy of the original Navy carrier-landing requirement.

The question of one engine versus two was settled for Burns during his tour with the 8th Tactical Fighter Wing in 1968-69. During that period 29 of his aircraft came back with one engine shot out; at the same time there was no instance of the problems of a damaged engine spreading to the good engine.

Burns recalls with pleasure the versatility that the Phantom conferred upon a commander. Aircraft could be switched from one mission to another as ground conditions dictated. During periods of relatively low MiG activity the Phantoms could be sent in with full bomb loads; if the MiGs surfaced (and the pilots prayed they would), the ordnance could be jettisoned and the F-4 was then a first-rate air combat fighter. From April to June 1967 the NVAF lost 32 aircraft and ceased combat for two months. During this period all Burns' F-4s were assigned to Alpha strikes in North Vietnam or Laos.

The F-4 brought the radar-guided missile into effective use and Burns notes that only the F-4 and the F-15 have used this type of weapon in combat (with the possible exception of the Mirage III and Foxbat). He points out also that there have never been any battles between a radar-equipped fighter and a non-radar type in which the exchange ratio favoured the latter. (Some commentators disagree with this, saying that F-4 intrusions into enemy airspace were often carried out with the radar switched off to avoid detection.)

The very versatility of the F-4 resulted in a lower exchange ratio and higher losses than otherwise would have occurred. When F-105 losses rose to an unacceptable level the F-4s took over the dangerous flak-suppression and anti-missile site missions. Some of the weapons the F-4 could deliver – the TV-guided Walleye glide bomb, for example – required techniques which made the aircraft very vulnerable to flak and SAM opposition.

Burns, unlike Steve Ritchie, feels that the Air Force would have done better to keep the GIB a pilot, but to give him the necessary training and upgrading prospects. This would have avoided the major morale problems suffered by the back-seaters.

In summary, Gen Burns feels certain that the Phantom

**Below: What it was like for Vietnam Phantom crews: endless days of rolling in over the countryside of North and South Vietnam, Laos and Cambodia. Here an RF-4C heads for its photo run.**

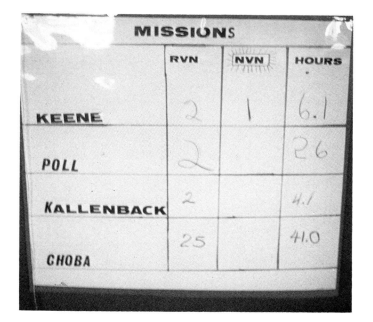

Above: Life on the ground for Vietnam Phantom crews contained echoes of earlier conflicts. Just as in the First World War, missions and combat time were logged on a board in the crew room. (Dr McGregor Poll)

Above right: The "Party Hootch" at Cam Ranh Bay. Left to right: Capt Dan Kallenback, Capt McGregor Poll, 1st Lt George F. Keene, pictured in April 1968. Squadron social life centred on the hootch (shack), and each unit (557th, 558th and 559th TFS) built one right outside its headquarters. Poll flew his first tour from Cam Ranh Bay between April 1968 and April 1969, then flew at Ubon from January 1970 to January 1971. He came back for a most unusual third tour from May to October 1972 at Udorn. (Dr McGregor Poll)

was the right aircraft at the right time in South-east Asia, and that if the North Vietnamese had had a similar type the air war over the North would have become impossible.

The GIB issue apart, Steve Ritchie's views on the Phantom are generally in agreement with those of Gen Burns. They both regarded the J79's smoke trail as a gift to the enemy. The bulky F-4, twice the size of the MiG-21, was even easier to acquire visually because of the long black streamers of smoke generated by the engines. Communist pilots were able to spot them from 20 or more miles away, particularly as they were given excellent GCI vectors.

What really riled Ritchie, however, was the poor performance and location of the single UHF radio. The UHF was absolutely essential in combat, yet the failure rate was abysmally high in the steamy atmosphere of South-east Asia. The radios usually failed just before a mission, causing an abort, or at the height of combat. When he briefed Gen Momyer, then head of Tactical Air Command, after his combat tour Ritchie asserted that the US should not spend another cent on new aircraft, radars, engines or missiles until it learned how to build and service a UHF radio. The location of the radio still brings a froth of fury to Ritchie's lips. It is positioned under the back seat and cannot be changed without pulling the ejection seat, with all the attendant delays and hazards of that process, and the ever present danger of incorrect re-installation. He recalls that after a heavy rain – a sometimes daily problem in South-east Asia – as many as 20-40 per cent of the UHF radios would be inoperative, resulting in lost sorties.

As much as he loves the aircraft that made him an ace, Ritchie also faults the layout of the cockpit. He feels that the stick was positioned far too low, requiring most pilots to reach out to use it, even with the seat adjusted. The Phantom was not as stable a platform as the F-105, nor as crisp in manoeuvre as the F-104. It was however docile enough to permit pilots to train the navigator GIB to fly formation, refuel and, when they felt very liberal, land the aircraft. This was a precaution against the time when incapacitation of the pilot might make it vitally necessary. Strictly against all local operating policies, it was good for the back-seater's morale and gave the aircraft commander a chance to start on the incredible amount of paperwork associated with every mission.

In analysing the opposition, Ritchie has a high regard for the fighters which opposed him. He considers the MiG-21's speed and turning ability to be excellent, but thinks that its weapon system was far inferior to the Phantom's. At 15,000ft and below the two aircraft were equally energy-manoeuvrable, although the communist pilots did not exploit this capability. The F-4 had an adverse-yaw problem near and past the maximum turn rate (19-20° angle of attack), and this was worsened if the pilot did not counter his natural tendency to use ailerons. The MiG-21 had roll rate troubles at maximum speed, as well as longitudinal stability problems.

Ritchie regards the MiG-19 as underrated; it turned even tighter than the MiG-21, and had good speed and acceleration. The later versions carried both missiles and 30mm cannon. Current USAF opinion still considers the MiG-19 to be a very tough opponent, particularly if it is equipped with heat-seeking missiles with a head-on capability. Many of the MiG-19s were supplied by the People's Republic of China, and there is some indication that these aircraft required more maintenance than their Russian-built counterparts.

The failure of the communist fliers to use energy manoeuvrability to the same degree as the Phantom pilots and their consistent failure to provide mutual support is considered by Ritchie to be extremely significant. He notes that on disengaging the North Vietnamese MiGs would invariably split, one sliding away in a 135° diving turn while the other chandelled up and out of the combat area.

Any comparison of Phantom and MiG must eventually dwell on the churning trails of smoke left behind by the American type's twin J79 engines. A flight of four Phantoms made enough smoke to betray itself from 25 miles away. Given the capability of the NVAF ground intercept system, the smoke was the final straw in taking away any element of surprise from an F-4 attack.

McDonnell engineers, in conjunction with the USAF, USN and General Electric, worked on the problem from the mid-1960s onwards. In 1967 McDonnell designed a system which automatically injected an Ethyl Corporation combustion improver (methyl cyclopentadienyl manganese tricarbonyl, CI-2) into the engine inlet manifolds. This was installed in all F-4Js from production block 37 onwards and retrofitted into all other Navy F-4Js and F-4Bs.

The Air Force declined to adopt the system, preferring to wait for the promised smokeless engines from General Electric, which in May 1972 submitted an engineering change proposal covering incorporation of the Long Life/Low Smoke Combustor system. The original system did not meet specifications, however, and it was not until 1976, far too late for Vietnam, that the Air Force finally gave technical approval for an improved system.

In combat a number of ad hoc methods were used to reduce smoke. The engines would be put in minimum burner, at which setting they did not smoke, and then a quick change of altitude would be carried out to throw off any

inquiring eyes. The Navy evolved a technique which was only slightly different. When being vectored towards known or suspected enemy aircraft, and before closing to visual range, one engine would be placed in idle (at which it did not smoke) and the lower stages of afterburner were selected on the other, which again did not smoke. This technique gave the desired airspeed for the approach and resulted in little yaw and only a slight increase in fuel consumption.

The other great Phantom bugbear was the "departure," a euphemism for going out of control and into a spin. It was coined in the early days of the Vought F-8 Crusader, which had even more severe problems in this respect, and has been used since to describe the loss of control which results when adverse yaw is introduced at a high angle of attack. Jack Krings, the personable and articulate director of flight operations at McDonnell, describes the departure problem of the F-4 as "pivotal," noting that it was eradicated in the F-15 and F-18 by improvements in controls, configuration and sensors.

**Left:** 1st Lt William Vasser suited up for a mission. He has a parachute harness, self-rescue tree-lowering device (good for 150 scary feet), 30 rounds of tracer ammunition, flares, radio, pin-gun flares, g-suit, extra batteries, water flask, knife case, radio, knife, maps and checklists, first-aid kit, 0.38-calibre pistol and a Mae West. *(William Vasser)*

**Below left:** Bill Vasser climbing into the Phantom that he christened *Nancy*. Vasser returned to fly O-2 light FAC aircraft after his F-4 tour. He was shot down and rescued in an area of North Vietnam where five crews had been lost in the previous month. *(William Vasser)*

**Below:** This F-4E of the 4th TFS, 366th TFW, based at Da Nang, made an emergency landing at Phu Cat in June 1971 after taking a 50-calibre shell in the right wing root. It remained at the base for 30 days for repair and flight test. After the war the aircraft was assigned to the 31st TFW at Homestead AFB, Florida, before being transferred to the Egyptian Air Force. *(Norman E. Taylor)*

The root of the problem was the Phantom's distinctive anhedral tailplane. Once the airflow over the tailplane began to rotate, as it would following a yaw input at high angle of attack, it persisted uncontrollably in that motion, forcing the aircraft into a spin. This tendency could have been countered if the F-4 had had two rudders (as do the F-15 and F-18), if the stabilators had been capable of differential operation, or if the anhedral had instead been dihedral. (In the original design anhedral had been applied as the solution to another aerodynamic problem.)

Some immediate solutions were available, and warning devices and intensive pilot training in recovery techniques reduced the number of departures to about six or eight a year. The problem does not arise in ordinary flight: it is most likely to occur in actual or simulated high-g air combat, in defensive manoeuvres, or in air-to-ground delivery manoeuvres, the last case being the most hazardous.

One of the most difficult problems associated with departure is the fact that the controls do not have the same effect when applied after the departure as they did before. At high angles of attack the ailerons, normally used for roll control, become big rudders, yawing the aircraft around and tending to turn it sideways. The rudder must now be used for roll control. The pilot who has just pulled up sharply and suddenly feels the aircraft sliding out of control finds this difficult to grasp. The situation is so confusing that the most effective recovery technique (quick application of rudder against roll, suppression of aileron input) is unlikely to occur to him. If the wrong technique is used (aileron against the roll), the situation may become irrecoverable. As a result some simpler techniques, which require a great deal of self-control but are more effective, have been evolved:

1 The pilot pushes the stick forward to reduce the angle of attack, and hopes that the aircraft comes out.
2 If it doesn't come out, the pilot pulls the landing drag chute in an attempt to force the nose down.
3 If this fails, and the aircraft is passing 10,000ft above ground level, the pilot ejects.

Below: *Betty Lou*, an F-4E of the 469th TFS, 388th TFW, seen at Korat RTAFB in 1968. Pilot was Col A.K. McDonald, back-seater Lt Jack Fisher. *(Albert Piccirillo)*

Above: An F-4D of the 480th TFS photographed on April 16, 1971. It is armed with six Mk IV 500lb bombs with 36in fuze extenders and four cluster bomb units. *(Norman E. Taylor)*

Below: A pair of 389th TFS F-Ds receive a final inspection from the "last chance" crew before departing Phu Cat on a mission over Vietnam on April 17, 1971. The aircraft are each loaded with six 500lb high-drag bombs and four napalm tanks. *(Norman E. Taylor)*

# 8. Phantom in the Middle East

The Phantom will forever be important to the people of Israel, for a wide variety of reasons. Foremost amongst these is the deep symbolic significance of the first Phantom delivery, on September 7, 1969. The War of Attrition was building up, President Nasser of Egypt and his Arab allies were openly planning to bleed Israel to death along its frontiers, and the long-established "romance with France" had come to an abrupt, mysterious halt when President de Gaulle of France had placed an embargo on arms deliveries to Israel. Fifty desperately needed Mirage 5J fighters, already paid for, were withheld at a time when the balance of apparent power had shifted heavily against Israel. The arrival of the Phantom indicated not only support from the US, and relief from dependence upon the caprices of French goodwill, but also provided the IDF/AF with a means of relieving the unremitting pressure against the Suez Canal front.

The Phantom went on to become the premier aircraft in the Israeli Air Force until it began to be superseded by the F-15 and F-16 in December 1976 and January 1980 respectively. By far the most sophisticated and expensive aircraft ever employed by the Israelis at the time of its acquisition, the Phantom also conferred upon its elite pilots a new measure of distinction. It was seen as demanding the ultimate in pilot skill if its full potential was to be realised. Israeli Phantom pilots were the *crème de la crème*, for fighter pilots have always been the most esteemed of Israeli warriors, and those who flew F-4s were known to be the best of the fliers.

The history of the Israeli Air Force (in Hebrew, *La Tsvah Haganah Le Israel/Heyl Ha'Avir*, or Air Force of the Israel Defence Forces) has been told many times. The term "born in battle" has been applied so often to the IAF that it has become a cliché, and yet no phrase better explains the use that Israel has made of the Phantom. The IAF came into existence in the heat of combat in 1947, equipped with a ragtag collection of Czech-built Avia S-199s (a stock Messerschmitt Bf109G fitted with a Junkers Jumo 211F engine), Austers, de Havilland Rapides, Piper Cubs and three tired B-17s. Even so, it was described by Prime Minister David Ben Gurion as the instrument which made victory possible in Israel's first war of survival.

The success of the new air force was due in part to the fact that it had dedicated volunteers flying for it. Though technically mercenaries, for the most part they earned so little money that this was a misnomer. They brought to the IAF what the Flying Tigers brought to China: a desire to fly and fight in a good cause, and they established standards of dedication and proficiency that are to this day held up as an example by Israeli leaders.

The IAF went on to a second total success against all odds in the 1956 war, primarily because of the skill and determination of its leader. A veteran of the Royal Air Force, Maj-Gen Dan Tolkovski, revolutionised the IAF from top to bottom, drilling ground crews and pilots with equal rigour, and insisting on the procurement of multi-purpose fighter-bombers as the basic Israeli air weapon.

By 1956, only three years after he became commander, the IAF had acquired North American P-51D Mustangs, de Havilland Mosquitoes, a single Boeing B-17, Gloster Meteors, Dassault Ouragans and Dassault Mystère IVAs in small numbers. The Arab air forces had been supplied with Soviet material in quantity, particularly MiG-15s, but the difference, as in every Middle Eastern war, lay in the quality of the pilots.

The much smaller Israeli Air Force eradicated the opposing air forces and proved to be decisive in winning the land battle. All of Tolkovski's efforts were justified in this classic campaign, which began on October 29, 1956, and ended in the muddle of the Anglo-French intervention and withdrawal on November 6. From the opening moments, when Israeli Mustangs destroyed Egyptian telephone lines,

**Below: Modern aircraft, ancient cities: the lower F-4s are almost lost in the background. For many Israelis the Phantom symbolises national survival.** *(MDC)*

Above: A cleaned-up F-4E makes a low pass over the runway. Here, as always, the Israeli Air Force censor has removed squadron insignia from the rudder, even though Israeli F-4 dispositions are well known to the Arabs.

Above right: The Phantom became a special aircraft to the Israeli Air Force and the Israeli people, offering them a strategic option for the first time. It is believed that at one time the IAF maintained a force of half a dozen Phantoms armed with nuclear weapons and kept on permanent alert.

Right: This crudely retouched photo shows where the censor blocked out numbers and unit identification, and an aircraft – possibly an F-15 – in the background.

to the end, when Nasser sought refuge in the Anglo-French embarrassment, the IAF had been supreme.

The lessons of the Suez campaign were quickly absorbed by the entire Israeli command, and IAF commander Ezer Weizman did not have any difficulty in procuring aircraft to match the MiG-19s and 21s which were flooding into the Arab air forces. Weizman engineered the build-up of a force of French fighters comprising 40 Ouragans, 60 Mystère IVs, 24 Super Mystères, 72 Mirage IIICJs and 25 Vautours.

Weizman was replaced as commander in 1965 by Maj-Gen Mordechai Hod, moving up to become head of operations and chief planner of the next major contest, the unbelievable Six-Day War.

Egypt, under President Nasser's guidance, had formed with Syria the United Arab Republic and by May 1967 had induced other Arab countries to contribute forces to those lined up on Israel's frontiers. The Arab world was gripped with a hysteria made up of the desire for revenge, the apparent certainty of victory and the traditional Islamic belief in the *jihad* or holy war. Some 650 Arab fighters, a mixture which ranged from Egyptian MiG-21s and Sukhoi Su-7s to Jordanian Hawker Hunters and Lockheed F-104s, were arrayed against the 196 Israeli jets. On the ground, the odds were even more formidable, with something like 500,000 troops and thousands of tanks against 50,000 regular Israeli troops, 120,000 reservists and perhaps 650 tanks.

Lt-Gen Yitzhak Rabin's answer, of course, was to launch Hod's air force in a pre-emptive strike, flying low under radar screens and attacking first the Egyptian Air Force and then in turn those of Jordan, Syria and Iraq.

The result was total victory. The IAF performed like an air force three times its size, with ground crews reducing turn-round times to seven minutes and pilots flying as many sorties as there were targets for. The war was won after the first three hours, and in the next six days an execution took place which left Israel in charge of Sinai and the Golan Heights but did nothing to reduce the deep hostilities in the region. In fact, as we shall see, the Arab world seemed to learn more from defeat than the Israelis did from victory.

The Israeli nation was delirious with joy at the extent of the victory. The armed forces, particularly the Air Force, were praised and rewarded, and certain problems which had become evident in the campaign were simply glossed over. On the map Israel seemed protected now as never before, with the vast Sinai as a buffer with Egypt, and the Syrians thrown back across difficult terrain.

Yet Israel did not have the manpower to occupy the territories it had won, and Arab resentment at the loss resulted in an immediate resumption of hostilities in the unspectacular but draining War of Attrition. The IAF Phantoms made their operational debut in this conflict, and in the view of some Israeli commentators they *lost* the war in the long run. Before we examine that war and subsequent Israeli employment of the Phantom, it would be useful to look at the IAF's selection and training process, and to compare F-4 experience in Vietnam and the Middle East.

The question "What makes a good fighter pilot?" has exercised some of the finest military minds since the birth of air power. There are hundreds, even thousands of books and articles dealing with the subject, ranging from hero-worshipping tomes like Raymond F. Toliver's *Fighter Aces*, via personal accounts like Billy Bishop's autobiographical *Winged Warfare*, to specialised reports like R. E. Doll's *Early aptitude-achievement discrepancies as predictors of later voluntary withdrawal from Naval aviation training*. The subjective approaches sometimes boil down to "tiger talk": an ace must have an instinct to kill, to go as close as possible to the enemy and knock him down from point-blank range. As von Richthofen said: "Anything else is rubbish." Other aces, however, speak of the need to be an excellent pilot, with full knowledge of the aircraft and systems, to know when to back off, and to be determined to a high but realistic degree. Given these attributes, all that is needed is timing and luck.

The Israeli Air Force is well aware of these views, but has

**Left:** A MiG-21 (centre) a MiG-17 (upper right) and an An-2 (lower left) at Gia Lam airfield, across the Red River from Hanoi. The picture was taken in 1967 by a USAF RF-101. *(USAF)*

**Above:** The extent of North Vietnamese preparations by 1967 is surprising: note MiG-17s in excellent revetments, and anti-aircraft emplacements. This picture of Hoa Lac airfield, near Hanoi, was taken in April 1967. *(USAF)*

of the American air effort in Vietnam, then the weapon's shaft was the enormous logistic chain of bases, ships, factories, depots, warehouses and administrative establishment that supported them. The cost per Phantom flying hour, though never actually accounted, was probably astronomical, for this huge back-up effort was in fact devoted to a comparatively small fleet of aircraft. The USAF maintained its F-4 forces at a relatively constant level, allowing them to decline during the reduction of ground forces, then expanding them rapidly when the North Vietnamese build-up began in 1972. US fighter forces did not build up on a scale comparable with that of the Second World War because the aircraft were immensely expensive and needed equally costly support in the form of tankers, ECM aircraft, and command and control. Though the Vietnam War represented a great drain on US resources, it never called forth a truly national

**Left:** A typical SA-2 site, with five missiles visible. Total area is about 700ft, with six launching pads. This site, located about 25 miles north-east of Hanoi, was photographed by an RF-101 in November 1968. *(USAF)*

**Right:** Three MiG-17s in revetments at a field west of Hanoi in October 1966. *(USAF)*

**Left:** Phuc Yen airfield in North Vietnam. Another 13 MiG-21s and 18 more MiG-15s and MiG-17s could be seen in other parts of the field. Photograph taken in mid-January 1967. *(USAF)*

Right: A RF-101 Voodoo took this picture of a North Vietnamese SAM on July 5, 1966. The SAMs were to become an ever more significant factor in the war. *(USAF)*

Below right: When the rules of engagement were relaxed the North Vietnamese airfields could finally be attacked. This is Queng Lang airfield after an attack by F-4s of the 8th TFW on June 14, 1972. Note the four large craters in the runway. *(USAF)*

Below: RF-4C reconnaissance crew from the 432nd TRW took this picture of destroyed and and damaged MiGs in revetments some three miles from the main runway at Phuc Yen in October 1967. The pilots were Maj Willis McDermitt and 1st Lt Michael Fenner. *(USAF)*

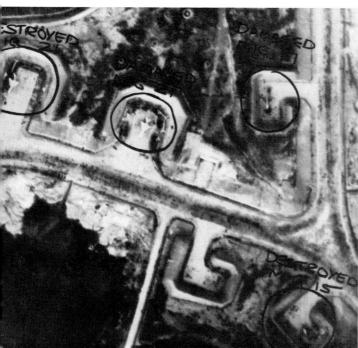

Right: Phantom crews, with smart bombs and much bravery, finally dropped the western span of the famed Thanh Hoa railway and highway bridge, located three miles north of Thanh Hoa City, North Vietnam, on May 13, 1972. The bridge was 540ft long and 56ft wide, and spanned the Song Ma River. It was a key link in the supply line from Hanoi to South Vietnam. *(USAF)*

Left: 480th TFS crew just back from a mission over Vietnam on April 16, 1971. Note squadron badge on gear door. *(Norman E. Taylor)*

Right: Two MiG kills decorate this F-4E of the 469th TFS, 388th TFW.

Far right: Another MiG-killer: F-4C 64-776 with three red stars. *(Copsey)*

Above: The one everybody wanted: "Last Mission Parade" for two members of the 480th TFS, 12th TFW. The lucky crew would taxi in behind a cavalcade of wing and squadron commanders, crew chiefs, maintenance personnel and friends. When they climbed out of the aircraft they were hosed down and congratulated with a bottle of champagne. This parade took place at Phu Cat in February 1970. *(Norman E. Taylor)*

Left: A typical 100-mission hosedown. The willing victim is from the 480th TFS. *(William Vasser)*

Right: A MiG-killer now with the Oregon Air National Guard. *(Copsey)*

effort of the kind required to give the armed forces everything needed for the job.

Gen Momyer notes that the nature of the targets demanded individual bombing (due in large part to the restrictive rules of engagement) and the number of aircraft that could be used on any target was limited. It turned out that a basic strike force going into the Hanoi area would consist of 16 bomb-carriers (for a long time F-105s, then Phantoms) with a mixture of ordnance best suited for the task. By the time the Phantom was available as a strike aircraft, smart bombs had also appeared, greatly increasing the effectiveness of the effort.

The strike group would be preceded by Iron Hand flak-suppression flights. These formations would usually consist of two aircraft carrying anti-radar missiles and two wingmen carrying conventional bombs. Early in the war the Shrike missile was used; it was then supplemented with the improved Standard Arm, which had more range. The Iron Hand flights would reach the target area about five minutes in advance of the strike force, hoping to catch the SAM batteries unawares. An intricate game developed, with the SAM batteries varying their tactics (simultaneous launches, optical launches with later radar pick-up, etc) and the Iron Hand units developing counters. It was a hazardous mission, for the Iron Hand aircraft had to precede the strike force, suppress flak during the strike, and then cover the bombers on their way out.

Later in the war chaff flights were introduced; F-4s or A-7s would attempt to mask the strike force from the SAMs with a cloud of aluminium foil. However, some crews felt that the chaff merely focused the attention of the enemy radars on the incoming strike force.

EB-66s provided stand-off jamming support in an attempt to prevent the enemy early-warning radars from vectoring MiGs against the strike force at long range. However, the powerful barrage jammers in themselves acted as a magnet, and after several EB-66s were shot down by MiG-21s it was decided to give these aircraft an F-4 escort. Thus each change bred further changes.

The entire strike force would be protected by a MiGCAP of Phantoms (BarCAP [barrier combat air patrol] or TarCAP [target combat air patrol] in the Navy), and the whole effort would then be assessed by flights of RF-4C reconnaissance aircraft.

The consensus seems to be that the US Navy developed the more flexible, responsive tactics during the war. Air Force pilots who flew the F-4 generally concede that their own effort might have been improved by the adoption of some of the Navy methods. The principal difference (apart from training philosophy and policy on crew integrity) lay in basic combat formations. The Air Force adopted the Fluid Four, a flight made up of two two-aircraft elements which provided mutual protection. Essentially the same formation was used by the Luftwaffe in the Second World War. The leader of the flight was intended to be the "shooter," and the Fluid Four was most suitable for turning engagements by aircraft equipped with guns. Thus it was far from perfect for the gunless, wide-turning Phantom until the introduction of the F-4E. The lead was separated from his wingman by 1,000-2,000ft to give some ease of manoeuvre, while the second element flew 3-4,000ft behind and 2,000ft above the first pair, weaving to provide coverage.

The second pair gave better visual and radar coverage, but they also had to work hard to keep the first element in view, so reducing their effectiveness. In actual practice skilled veterans usually modified the Fluid Four into a pair of Navy-style Loose Deuce flights, allowing more independence and tactical flexibility, and providing more shooting opportunities. Loose Deuce was picked up by the Air Force after the war, and has since undergone further modification into Fluid Two and Fluid Three. The Navy itself is now reconsidering Loose Deuce, but during Vietnam it was gospel.

Loose Deuce comprises a pair of aircraft flying abeam and one to two miles apart. The wingman strives to maintain the abeam position, varying altitude as necessary and reducing the separation when visibility falls. The wider spread makes visual search easier but also puts a premium on training, quick comprehension of radio transmissions, and the use of yo-yo manoeuvres to maintain the formation. The two F-4s will always seek to engage one enemy; two on one is essential if the combat initiative is to be preserved when opposed by a more manoeuvrable aircraft.

There is one significant difference from Fluid Four, numbers of aircraft apart: the Loose Deuce tactical lead passes from one aircraft to the other, depending upon who has the radar or visual contact. Whoever gains the contact becomes the shooter, while the other aircraft is the free fighter, giving support, cutting off the enemy, preventing surprise attacks, and in general keeping the pressure on.

The following two accounts from the Joint Munitions Effectiveness Manual illustrate how US tactics improved as the war went on. The first case shows how in 1965 MiGs were able to mix it up with a USAF unit and inflict some losses; the second, in 1972, shows how sophisticated the USAF defence methods had become, and how dangerous life was for the MiGs.

"On April 4, 1965, the Air Force went against the Thanh Hoa Bridge. Four F-100s flew MiGCAP for 48 Thuds. (Lead was a Second World War-experienced squadron CO and all the Hun drivers had lots of tactical experience.) Everybody knew about the Navy's MiG engagement yesterday [MiG-17s had attacked Navy F-8s and A-4s, with no decisive results] at the same place. In spite of all the above, everyone was thumping around at 300-350kt.

"The No 2 F-100 MiGCAP pilot saw two MiG-17s at 9 o'clock low, three miles, and called them, but Lead thought he was referring to some distant Navy F-8s and responded 'Negative, friendlies'. No 2 reiterated that they really were MiGs while he was looking through his cards for the MiG alert codeword.

"At this point Lead said 'Let's have at them' and the F-100s lit burner and dove for the MiGs. It took a bit before they started closing on the MiGs due to their low initial speed.

"The MiGs were heading for our F-105s that were orbiting at – you guessed it – 325kt. The third F-105 saw them, identified them after one or two seconds, and called an urgent break. The Lead and No 2 F-105 continued on. The MiGs flashed past No 3 and No 4 and opened fire on the lead Thud, hitting him. The second MiG opened fire on No 2, who transmitted: 'Lead, you have a MiG behind you . . .

Above: But it wasn't all hosedowns and red stars for the Phantom crews: a 37mm anti-aircraft round destroyed one engine of F-4D 66-0275 and the aircraft force-landed at Udorn RTAFB. The incident took place in 1968. *(Albert Piccirillo)*

Right: The hole caused by a 37mm hit on Fred Olmsted's F-4. The big shell blew out the left engine and caused a massive loss of hydraulic fluid. *(Fred Olmsted)*

[pause] . . . I've been hit.'

"The MiGCAP F-100s were now in AIM-9 range. Lead F-100 hesitated, worried by the proximity of the MiG-17 to the F-105. He fired just as No 2 called: 'I've got my pipper on a MiG, am I cleared to fire?'

"Lead's missile passed over the MiG and he cleared No 2 to fire while he pressed in for a gun shot. The MiGs broke upward and Lead lost them. No 2 went up with the second MiG but couldn't match his turn. The MiG continued hard over the top, down, and started up again. No 2 did something akin to a high yo-yo and when the MiG pulled up again No 2 turned right and down and cut across. He put his pipper on the target and fired 280 rounds, closing from 1,500 to 800ft, but without observing hits. No 2 passed over the MiG, belly to belly, hit his jetwash, and banked. The MiG went down into the cloud deck and was gone.

"While all this was transpiring, four ResCAP F-100s also had a fight. When the MiGCAP called the MiGs the ResCAP tallied a bogey dead ahead on a reciprocal course. As it flashed by and rolled into them, it was positively ID'ed as a MiG-17. This MiG swung in behind the second element of ResCAP F-100s while another MiG turned in behind the first element. Like everybody else, the ResCAP F-100s were in orbit at 300-325kt.

"The lead F-100 called a break to No 2, who lit burner and pulled hard. He saw the MiG's nose was already low so he raised his and held g to increase the overshoot. The MiG slid in behind Lead, who called: 'He's on me now.' No 2 selected radar sight and did a nose-high rudder reversal, uncaged the sight, set the wingspan at 32ft, came out of burner and dove back. The MiG was in position for a good shot so No 2 pulled lead and fired a long burst. The MiG reversed immediately and then back again in a sort of a split S. Lead, noting that his speed had bled down to about 250kt, lowered his nose to accelerate; at about this point No 2 lost him. No 2 went into a nose-high roll to the left, then straight down, and added power to close on the MiG. He couldn't track the MiG so he caged the sight, fired (no observed results), put the pipper on the MiG tailpipe, fired a long burst (observed a flash), and started a hard pullout. Going idle and boards [cutting power to idle and selecting airbrake] and almost blacking out brought No 2 straight and level at wavetop height about 1½ miles offshore.

"The second MiG fired a 90° deflection shot at No 3, then pulled behind No 4, who was able to dive away from him, after which the MiG disappeared into the clag.

"The two F-105s didn't make it back, and their pilots weren't recovered."

Seven years later a lot had been learned. The Manual goes on: "On August 19, 1972, Pistol, a flight of four F-4Es, was performing chaff-escort duty in support of a Linebacker strike near Kep Airfield. The chaff force refuelled feet-wet and ingressed over Hon Gay (090° and 80nm from Bullseye), preceded by the ingress CAP flight. The weather east of Hanoi was exceptionally clear. Crossing the coast at 15,000ft, Pistol 01 (4th TFS, 366th TFW) called for jettison of the centreline tanks and initiation of the prebriefed element-weave formation two miles in trail with the chaffers. As Pistol weaved, Red Crown advised that bandits were airborne at Phuc Yen and proceeding east.

"In Pistol 03's (Capt Sammy C. White) words: 'Red Crown was really good that day. The first call we got was

**Below:** An F-4C being recovered after crash-landing on the runway at Bien Hoa, South Vietnam, in February 1966. *(USAF)*

**Above right:** A Phantom sits forlornly on the sidelines at Da Nang, its gear driven through the wing during a hard landing. *(William Vasser)*

"Bandits, 098° for 47nm from Bullseye". Red Crown continued giving calls as the bandits tracked outbound towards our position. My back-seater (Capt Frank J. Bettine) had Bullseye dialled in the inertial navigation system (INS), and as Red Crown called the bandits' position he would compare that to our INS and was able to tell me just where they were in relation to us. It was apparent from the calls that the MiGs were using a parallel head-on intercept, planning a left turn to our westerly heading.'

"Approaching a point 080° and 50nm from Bullseye, Pistol's elements were on a converging course in the weave. The lead element (01/02) was on the right (north) side of the formation in a left turn. As the elements crossed, 02 relates: 'My back-seater (Capt Forrest Penney) picked up the lead MiG (a MiG-21) attacking from 6 o'clock and called for our element to come hard left. Our element leader reasoned that if the MiG was close enough to see he was close enough to shoot, so he called "Break left." I was on his left and I lost sight of him in the break.'

"Halfway through the turn, 01 saw MiG 1 overshoot to the outside of the turn and called for the element to reverse back to the right. However, 02, who did not have 01 in sight, continued his break and became separated from the flight.

"When 02 called the MiG at 6 o'clock, 03/04 were just crossing over 01/02. Spotting MiG 1 at his 4-5 o'clock, 03 took his element into a hard, nose-high turn. His comments: 'The MiG looked like he was going to follow the lead element in their left-hand break. Then he rolled wings-level and looked like he was going to come with our element. He was evidently surveying what was going on. He must have thought "I don't have much of a shot at anybody." Anyway he decided to disengage.'

"Another MiG-21 (MiG 2) had been in trail with MiG 1 for the intercept. Observing MiG 1's attack fail, MiG 2 broke off to the south and disengaged.

"MiG 1 dropped his nose to accelerate and rolled off to the right, crossing under 03's element in their climbing right turn. Rolling over the top to the right in a barrel-roll manoeuvre, 03 rolled out at MiG 1's 7 o'clock approximately 8,000ft back. MiG 1 then pulled his nose up about 45° as 03/04 pursued. The MiG then rolled to the left and began a 20° descent to 03's 12 o'clock; range was still 8,000ft.

"White describes the ensuing action: 'The switches were set up for a manoeuvring radar lock-on, 5-mile boresight, high g, short pulse. I put him in the pipper and tried an auto-acq [auto-acquisition]. The thing locked onto him on the first try. Frank Bettine called: "You're locked on. Thousand one, thousand two, thousand three, thousand four; you're cleared to fire!" During this the MiG made an easy roll to about a 45° left bank. When my GIB said "Thousand four, you're cleared to fire" I pulled some lead and squeezed the trigger. About a "week and a half" later the missile came off! Is it 1.2sec they advertise from trigger squeeze to the missile leaving the rail? In that 1.2sec I checked my switches twice, kept one eye on the MiG, and was even pumping the stick trying to shake the damned missile off! I was just getting ready to kick myself all over NVN for having my switches set wrong when all of a sudden this freight train rode up from under me on the left side. It felt so good I did it again, and the missile on the right side came off. The first missile hit him right in the tail. I would say the angle off was no more than 20° at the time. As he came out of the fireball I saw another quick explosion in the engine section. He trailed white smoke for about 3,000ft, the canopy came off, and the pilot ejected. Then we [Pistol 01/03/04] made our turn and egressed back toward Haiphong.'

"Pistol 02, meanwhile, was on his own. He recalls: 'After the break my back-seater called off a MiG about 2 o'clock. I looked up and there he was. As it turns out this was the wingman [MiG 2]. He was headed south away from the engagement. Had I realised then that the one we broke for was gone, I could have gotten him easily. But this was my first time up there and I was pretty scared. So I punched tanks and started S-ing out towards the coast, picking up about Mach 1.2 at around 15,000ft. When I got feet wet we all rejoined.'

"As Pistol began egress, the ingress CAP flight received vectors from Disco against MiG 2. At 15nm Lead obtained a full-system lock-on and attempted to close in a tail chase. They were not able to close until the MiG slowed for a landing approach at Phuc Yen. As the ingress CAP closed to missile range, MiG 2 broke into them. Lead fired two AIM-7s; they were out of parameters and both missiles impacted the ground. The flight then egressed to the west."

What a difference: from amateur flubbing in 1965 to closely co-ordinated, well developed, mutually protective tactics in 1972.

*Above:* The end of an incredible saga. This 433rd TFS, 8th TFW, F-4D (66-0249) was extensively damaged by AAA fire. The entire radar section was blown off, resulting in an enormous amount of flat-plate drag. The weapon systems officer ejected but the pilot brought the aircraft back to Ubon and made a wheels-up landing. Note the MiG-kill symbol and the extended ejection-seat rail. Few people believed that the aircraft could have flown in such a condition. *(Albert Piccirillo)*

*Left:* Battle damage to the tail of F-4D 66-0264 of the 166th TFS, 354th TFW. *(Albert Piccirillo)*

*Below:* Unusual twosome: a Cambodian MiG-17 and a 389th F-4D on the Phu Cat flight line. The MiG was flown to Phu Cat for testing by a team from Wright Patterson AFB, Ohio. It was later escorted back to Cambodia by two 12th TFW F-4Ds. The picture was taken in December 1970; in January 1971 the MiG was destroyed on the flight line at Phnom Penh, Cambodia, by Vietcong sappers. *(Norman E. Taylor)*

yet more factors to take into account. The country is small and its resources are limited; both aircraft and pilots must be husbanded. In any conflict Israel will always be vastly outnumbered by the combined Arab forces, and it is hard to envisage any political settlement which would guarantee that the Arab world would not suddenly unite against the Israelis. The resulting IAF attitudes strongly influence its training philosophy. In war the IAF expects to do the job required of it, regardless of how bad the odds, by "finding a way". This means that no matter what new enemy fighters, missiles or tactics are introduced, the personnel of the IAF will nonetheless prevail in the heat of battle. Maj-Gen Ezer Weizman set the tone when commander of the Air Force by writing: "Numbers don't count; only effective missions do. In the age of sophisticated weapons, we try to make the man in the cockpit count above everything."

Under Gen Mordechai Hod, Weizman's ideas on the importance of the individual were further defined. Two IAF aircraft, a leader and a wingman, are sufficient for any aerial battle, no matter how many of the enemy are involved. For those two aircraft will be piloted by men who have passed through the most rigorous training in the world, and who may have engaged in as many as 400 to 500 combat sorties per year.

In Israel universal military training begins for all citizens at the age of 17½. If a person volunteers, has had ten years of schooling, is medically fit and has an average IQ or better, he can enter the pilot training process. Induction is at the age of 18, after which a series of tests are given to determine aptitude. These are at first simple pen-and-paper tests of mechanical comprehension and eye/hand co-ordination, with some simple mechanical assembly tests to screen out the obvious misfits. Then comes a much more intensive screening at the Aviation Research Medical Facility, where additional tests, including personality assessments, are given.

There follows an intensive ten-day "tent city" during which the surviving candidates are subjected to physical and mental stress by such means as obstacle courses and group activities. During this period the candidates are closely scrutinised by both pilots and clinical psychologists. At the end of this there is another screening, this time based upon observed general motivation, scores relating to aptitude for officer training, and two sociometric scales, the "pilot" and "friend" scales. These ratings are then combined with the earlier scores and the final selections made. Great emphasis is placed throughout on peer acceptance, for the Israeli Air Force demands that its aircrew be utterly dependable in protecting the interests of their friends and comrades. At this point approximately 50 per cent of the candidates have been washed out and returned to other military duties.

Next comes four months of ground school and further

selection based on ten flights in a Piper PA-18, followed by four months of infantry training before entering one of three programmes. The candidate who has endured to this point will be routed into the fighter stream, on which he will receive four months of primary and four months of basic flight tuition in the dependable Fouga Magister, or sent to helicopter or navigator training.

All students then do four months' work in academic subjects at the University of Beersheba before being sifted once more: the most successful candidates go to four months' advanced training in the A-4 Skyhawk, while those with less aptitude for fighters go to helicopters or Dornier Do27s. Navigators continue their own course.

The fighter pilots who successfully complete their course go on to 18 months of operational work in A-4s before being upgraded to the Eagle, Phantom, Mirage or Kfir. There is a 20 per cent washout rate during the Piper training, and 10 per cent cuts during primary and advanced training in Magisters. So it is a very select group indeed that make their way into the operational A-4 squadrons, where once more they undergo a refining process designed to select those who will go on to the Phantom units.

Operational pilots hone their proficiency with 400 to 500 sorties per year, a figure that staggers US pilots, who feel fortunate when budget and other conditions permit 100 sorties a year. Still more depressing was the plight of the Luftwaffe with its F-104s, whose pilots were expected to master the worst weather in Europe on perhaps 50 sorties per year.

When Israel's Phantoms arrived in 1969, however, they were parcelled out carefully to pilots who were highly experienced and had already had training on F-4s in the United States. The F-4s arrived at a time when Egypt's strategy in the War of Attrition seemed to be paying off. The pressure on the Suez Canal line and upon the border with Syria, the daily losses due to shelling, and the steady reinforcement of the Egyptian missile and flak batteries all frustrated the Israeli leadership.

In many ways Israeli use of the Phantom has been a mirror image of American experience in Vietnam. Israel's F-4s have had to range outwards from home bases to strike targets in Egypt, Syria and elsewhere. In Vietnam they flew from Vietnamese and Thai bases and from aircraft carriers to hit targets in South and North Vietnam. In the Middle East the Phantom crews and their comrades face a host of enemies; in Vietnam they outnumbered the opposition. In Israel the Phantoms are operated by a nation with an iron will and unanimity of purpose; in Vietnam they were hampered by the irresolution of the American leadership and the resultant inhibiting rules of engagement.

In both theatres the Phantom crews were much better prepared than their adversaries, although Israeli crews ultimately came to be better trained and integrated than their American counterparts in Vietnam. Finally, in both the Middle East and Vietnam the F-4 was the premier fighter of its time, able to perform more roles more effectively than any other type.

The War of Attrition began within three weeks of the end of the Six-Day War; it would continue until August 1970, when a precarious ceasefire between Israel and Egypt was arranged. In those three years the area around the Suez

Canal had become a continuing testing ground for the two superpowers, with the Soviet Union pouring missile and anti-aircraft defences into the area while Israel systematically improved its air force with US equipment. By 1968 Egypt had increased its forces to more than 150,000 troops along the Canal, and was trying to erase the profound psychological effects of the defeat in the Six-Day War. The Soviets had supplied massive amounts of artillery as well as anti-aircraft equipment, and a devastating rain of fire was poured on the Israeli positions.

Israel's chief of staff, Maj-Gen Haim Bar Lev, had ordered the construction of a series of strongpoints along the Canal, and a huge sand barrier which was intended to prevent amphibious vehicles from launching a surprise attack. It was also decided to begin a series of attacks aimed at targets well behind the Canal line and deep in Egyptian territory. Those thrusts were designed to convince the Egyptians that the war was not being fought only on the Canal, but that they were vulnerable everywhere. This was the war that the Phantoms won – but lost.

The first significant step in this phase of operations was a brilliant Israeli commando raid which penetrated to the heart of Egypt, attacking bridges and electric transformer stations 220 miles from the Israeli border. Then, at the end of 1969, the first Phantom squadron became operational, and on January 7, 1970, the F-4s attacked an SA-2 missile site at Dahasur, 21 miles south of Cairo. There followed attacks on supply depots within Egypt, and some costly strikes on surface-to-air missile sites, operations which did not bode well for the future. There were, inevitably, cases in which bombs went awry and civilian areas, in one instance a primary school, were hit.

The level of effort was high; in the first four months of 1970 the Israelis flew 3,300 sorties and dropped 8,000 tons of bombs on Egyptian targets, knocking out 80 per cent of the Egyptian defences. The Phantoms were doing what was asked of them, and more, but ironically this led to even bigger problems for the Israelis. Nasser went to the Soviet Union and, according to reliable reports, went down on his knees to beg for assistance. It came, in massive quantities and displaying an unprecedented degree of modernity.

The Soviet presence rose to more than 15,000 troops; entire air bases were manned by Soviet personnel, as were SAM batteries and radar defences. The MiG-21MF (Nato code name Fishbed-J) began to appear alongside the standard Egyptian MiG-21F and PF (Fishbed-C and D), and it was flown by Russian pilots. The Israelis, wishing to avoid further confrontation, began to refrain from combat with the Russian pilots, and halted their deep-penetration

**Far left:** Israeli F-4E with refuelling boom extended. The Phantom made such an impression on the Egyptians that President Anwar Sadat singled it out for praise in talks following the Yom Kippur War. *(MDC)*

**Left:** The AN/ALE-40 chaff dispenser can be seen at the aft end of the wing pylon on this Israeli F-4E. *(Lon Nordeen)*

**Below:** Shots of specific sites on Israeli airfields are very rare. Israeli crewmembers, ground and air, work under spartan conditions. This aircraft is said to be assigned to 119 Squadron, and to have accounted for seven MiGs. It carries the identification number 109. *(MDC)*

raids into Egypt. The Soviet pilots showed similar discretion.

The Egyptians immediately renewed their efforts along the Suez Canal, with massive artillery and air attacks. New SA-2 sites were built, and manned by joint Russian and Egyptian crews. On the night of June 29, the Soviet/Egyptian forces moved in 12 improved SA-2 and three SA-3 batteries, along with additional anti-aircraft protection. On the 30th two F-4Es were shot down by the new defences, and in revenge the Israelis started a massive fighter-bomber attack against the missile batteries. It was not effective: the new defences had gained a measure of control of the air that they had never before enjoyed.

On July 30 came an air battle that overjoyed the Egyptian pilots despite the fact that the aircraft going down in flames were all MiG-21s. The Russian instructors had been tongue-lashing their Egyptian students, criticising their tactics and combat skills, and accusing them of lack of aggressiveness. On the 25th MiG-21s had attacked two Israeli A-4s, damaging one, and Russian confidence seemed to have risen as a result. On the 30th a few Skyhawks made a strike against the Nile Valley. The Russian-manned Fishbed-Js swept in to intercept, only to be sandwiched by Mirage IIICs from above and Phantoms from below. Four MiG-21s were shot down for no losses to the Israelis; that night Egyptian officers' messes resounded with sardonic cheers.

The air war was almost inevitably a source of satisfaction for the Israeli pilots, despite the fact that Egyptian skill and courage had improved enormously since the Six-Day War. But in the battle against the missiles things were different, calling for every ounce of skill and motivation from the Israeli crews. The following account appeared in the *Israeli Air Force Journal* of October 1981. Taken from an article entitled "Motivation That Rises from the Ground," it is a combat narrative by Brig-Gen A, as told to Hanah Zamar: "Towards the end of the War of Attrition, in July 1970, came the famous attack on the missile defences on the Canal, which coincided with the rolling forward of the missile screen by the Egyptians. On July 17 we attacked five missile batteries that came within 20 miles of the Canal and threatened to paralyse the ability of the Israeli Air Force to attack the enemy's ground deployment along the Canal. In this attack I participated as the leader of a formation whose mission it was to attack one of these batteries.

"In front of me was a formation from another squadron, led by Lt-Col Hetz,* who was shot down. I saw him hit. He was a number of miles in front of me. I saw the missile battery that fired at him. They also fired at us, around 50 to 60 missiles which we evaded safely. I wasn't very worried. I began an attack on the battery that had fired at Hetz. As I approached the battery I saw a missile being launched at me. I told my No 2 to be prepared, that they were shooting at us

---

*Lt-Col Hetz was Lt-Col Shmuel Hetz, deputy commander of an Ouragan squadron during the Six-Day War and first commander of Israel's first Phantom squadron. He was killed on this mission and his navigator, Menachem Eini, was captured. Eini was later appointed Israeli air attaché in Washington.

**Right:** Israeli F-4E with munitions displayed at an open-house event. *(MDC)*

**Far right:** On-board camera reveals the centreline gun and retarded bombs dropping away as an Israeli F-4 speeds over the target. *(MDC)*

**Below:** F-4E laden with a variety of iron bombs. Aircrew weapon skills are honed in the course of 20 to 40 missions a month. *(MDC)*

from the battery at 10 o'clock. He acknowledged that he saw the battery and the missile. We broke to fly towards the battery, to attack it, as the missile came towards us.

"Until this time all the missiles that were fired at us were of the SA-2 type, very large and easy to see from a distance; in Vietnam the Americans called them 'flying telephone poles'. The routine was to break hard when you saw the missile at its normal size; the missile would not be able to follow the turn, and would fly on harmlessly.

"This time they fired a new missile of the SA-3 type, which is smaller and more manoeuvrable. As expected, I saw the missile coming towards me. I saw that when I dove it dove and was drawing a lead to intercept me. I decided to wait until I could see the missile large enough and then I would break to shake it off. But this missile was smaller than the SA-2 of course, and so the break was rather late. It managed to save me from a direct hit but the missile exploded very close to me. The aircraft was entirely sprayed with shrapnel: we found about a thousand holes, some of them right through the cockpit.

"I was able to look out of the Phantom's cockpit through the holes. Some of them were around me and there were tears in my flight suit, but I wasn't really wounded. The aircraft had been hit very hard, however. I said to myself that I would make every effort to avoid being captured. I succeeded in turning the aircraft homewards, on one engine with the other one on fire, at the lowest possible speed, and with a number of missiles pursuing us and AAA filling the sky.

"My No 2 saw the hit but he still had his bombs and decided that he could not jettison his bombs to help me; he continued the attack, bombed the battery and then rejoined me, flying nearby. I didn't have any radio communication with him; I couldn't even speak to my navigator,* who was sitting behind me. Communicating with signs, I said that we were all right. I had to jettison the armament to make the aircraft lighter. The aircraft jumped as if it had been hit by a missile. The navigator thought that we had been hit again, so

---

*Brig-Gen A's navigator on this mission was Maj Shaul Levy, who was killed on October 7, 1973, during a strike on the Syrian missile defences while flying in a Phantom piloted by Lt-Col Ehod Hankin.

I gave him the sign that everything was okay, we were still flying.

"We managed somehow or other to cross the Canal, flying with all the cockpit lights lit up and one engine on fire. We decided to fly on and land at Refidim, some 80 kilometres from the Canal. I don't known how, but we succeeded.

"As we lowered the wheels I saw that we couldn't lower the flaps. I also saw that the aircraft couldn't fly straight at less than 400km/hr, about 220kt. It was plain that the landing would have to be made at this speed. We landed believing that we still had the emergency brakes, the only remaining hydraulic system. But then we found out that this too had been hit. So there was nothing to stop the aircraft on the ground. We touched down at 300-400km/hr and the aircraft ran on down the runway at this speed while we sat there, unable to do anything to stop it or keep it on the runway. The mistake we made was not to eject at this stage. This was one of the hardest decisions I've ever made, but after coming all this way to eject now would have been like leaving a friend in trouble.

"We left the runway at a speed of 200km/hr and the aircraft ran another kilometre on the ground, passing within a metre or two of ditches, walls, damaged aircraft (MiGs abandoned at Refidim), cannon, AAA, all sorts of things. Finally the aircraft stopped on a small hill or pile of dirt. The nosewheel broke. We climbed out and saw that we had been even luckier than we thought. When we jettisoned the armament we had forgotten about the missiles under the fuselage. The engine fire had continued to burn and had destroyed the structure around one of the missiles, melting its warhead. It had melted drop by drop as the aircraft ran along the ground."

The rapidly escalating conflict was halted by a ceasefire in August 1972. The Phantoms had won in that they had forced the Egyptians to abandon their attrition tactics on the Suez Canal, at least for a while. But they had failed to solve the surface-to-air missile problem, and they had induced Nasser to invite the Russians to intervene on a massive scale. In spite of their losses the Phantoms had secured a tactical victory, but they had inadvertently brought about a strategic reverse.

The Israeli Air Force attained an exchange ratio of 25 to 1 in air combat, claiming 100 Arab fighters destroyed for the loss of only four Israeli aircraft. But on the ground the story was different. Arab sources claimed more than 300 Israeli aircraft shot down, including nine F-4Es lost to SAMs; this is perhaps an exaggeration but is at least indicative of the effectiveness of a combination of large numbers of missile batteries and radar-directed anti-aircraft cannon.

Gen John Burns, in reflecting on the Israeli effort, speculated that their preoccupation with the air-to-air exchange ratio may have diverted resources from the campaign to eradicate the surface-to-air missile threat. This omission was to be almost fatal in the next Middle East war.

President Anwar Sadat succeeded Nasser after the latter's unexpected death in 1970. He was widely assumed to be an interim figure, a safe rein-holder until someone else seized power. Instead he proved to be the most effective of all the Arab leaders, in both war and peace. Sadat allowed the Russian build-up to continue until it finally touched upon the sensitivities of the Egyptian people and the security of his own regime. Then in August 1972 he expelled the Russians while somehow managing to retain grudging resupply support.

Sadat then began an intensive training programme which was also part of a giant deception plan. The Egyptians planned to build up impenetrable missile defences along the Suez Canal, while Syria did the same along its border. Egyptian troops exercised continuously, making feints across the Canal, and maintained a level of alert that in time began to lull the Israelis into lowering their guard slightly.

The Egyptian plan worked, but only because the Israelis inadvertently collaborated in it. Sadat had counted on the Israelis' eventually realising that a build-up was under way and launching a pre-emptive attack, only to spear themselves on the deadly SAM and flak belts. Instead, Israeli intelligence discounted the build-up in late September and early October because it did not fit their preconceptions of what should be happening.

As a result, when the Egyptian and Syrian forces attacked at 1400hr on October 6, 1973, the very moment that Israel was beginning the Yom Kippur national holiday, the IDF was in disarray. Because of a political decision, mobilisation had not started, and only got under way at 2 pm, when the orders were finally given. Immensely satisfying for the Arab pilots, it shook Israel to its foundations, for all of the conditions of the glorious Six-Day War now seemed to be reversed.

The following account by Lt-Col N in the *Israeli Air Force Journal* of June 1980 tells how it appeared to the Phantom pilots: "On the afternoon of Yom Kippur we knew that this day would not end quietly. Close to 1400hr, with all the aircraft armed and prepared, the controller reported to me that there were tens of targets flying in low over the sea. I told my No 2, 'This is it, I am taking off', even though I had no orders.

"The sirens were sounding when the aircraft left the runway and began to gain height. I looked back to see my No 2, and there were smoke columns rising from explosions on the runway.'This is really war', I thought.

"I pulled up hard, releasing my tanks, and made a pass at a MiG-17 just pulling up from his bombing run. The Egyptian dove, but my missile impacted and he blew up.

"Now the sky was full of aircraft; three MiGs tried to get on my tail, bursts from their cannon coming past. I broke hard and one MiG disengaged while I pressed the attack on the other two. They dropped to 150ft, flying under antennae

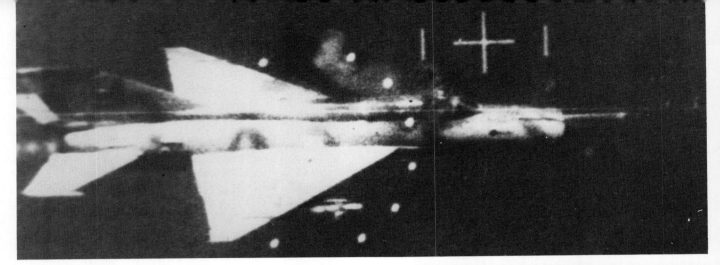

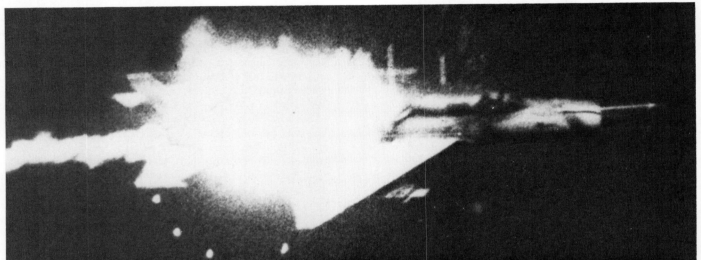

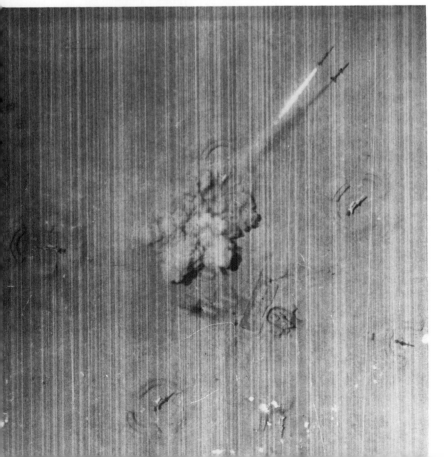

**Top:** Israeli Phantom crews relish the challenge of air combat. Here a Syrian MiG-21 takes cannon hits during one of the rare air-to-air gunfire duels of the Yom Kippur War. *(Aviation Week and Space Technology)*

**Above:** The MiG-21's internal fuel explodes as a result of further cannon strikes. *(Aviation Week and Space Technology)*

**Left:** The Egyptians hit back: an SA-2 streaks off its launcher, leaving behind a miniature dust storm. The Egyptian and Syrian SAMs proved to be extremely effective in the War of Attrition. The Israelis learned quickly, however, and countered them with relative ease in the 1982 Lebanon incursion.

**Right:** Egyptian MiG-21 gun-camera film showing attacks on bomb-carrying Phantoms during 1973 strikes on Nile Delta airfields and Port Said. *(Aviation Week and Space Technology)*

and electric wires; one MiG broke hard and disappeared into the sea. The remaining MiG was tougher; the pilot pulled high g and I stayed with him. Suddenly my left engine cut out; I began to lose speed and height, and was almost on the deck before I got a relight. A MiG came in on my tail, cannon firing; I reversed and came behind him, firing my own cannon. No hits. Then I put my last missile to him and he crashed on the shoreline."

In the border areas along the Canal and on the Golan Heights bombs and shells were falling on a nation which was not unprepared for war but which had been astonished by the suddenness of the onslaught. Israel's life depends upon vigilance, and she had been asleep.

The responsiblity for repairing the damage while the rest of the forces were mobilised fell upon the Israeli Air Force. The first task was to halt the advancing Egyptian and Syrian forces, acting in place of artillery and tanks until the Israeli ground forces were in place. Then the Israeli Air Force shifted over to the offensive, attacking airfields in Egypt and carrying out, for the first time, a strategic air offensive against Syria. The commander of the IAF, Maj-Gen Benjamin Peled, stated later that he had three objectives: first, to shift the centre of gravity of the fighting from the Egyptian to the Syrian front; second, to overcome the almost impenetrable missile defences; and third, by achieving the first two goals, to prevent the Jordanians from entering the war.

The switch of effort to the northern front was designed to place the Syrians in such danger that the Egyptians would be forced to move out from under their missile umbrella. At the same time, the Israeli Air Force paid dearly in losses to flak and SAMs while it blasted a lane through the missile defences, destroying a central command and control post. The contest at first appeared unequal: in the first few days of the war the combination of the relatively new SA-6 Gainful and SA-7 Strela and the older SA-2 and SA-3 with formidable anti-aircraft artillery (primarily the radar-directed ZSU-23/4) seemed unbeatable. The Arabs fired missiles at a prodigious rate, so fast in fact that their Soviet suppliers were unable to meet demand.

But Israeli determination to "find a way" finally prevailed. By the fourth day of the war events were beginning to take a familiar turn. By the eleventh the Israelis were crossing the Canal, and by the seventeenth (October 23) they were a threat to Cairo. The deteriorating situation on the ground had forced the Arab air forces into combat, and the IAF had achieved a massive victory. More than 550 Arab aircraft were destroyed, and the Israelis enjoyed complete air superiority over the battlefront, now only sporadically defended by SAMs. Israel is reputed to have lost only five aircraft in air-to-air combat, but about 100 to ground fire.

Once again the Soviet Union and the United States intervened and established a ceasefire. Russia did not wish to see the client states that she had armed go down to total defeat; the United States felt that the situation was favourable enough without risking an actual Soviet intervention with the airborne troops that had been mobilised.

The Israeli Air Force, with Phantoms leading the way, had held off enemy ground forces for long enough to permit the Israeli Army to mobilise; it had defeated two tremendous SAM and flak defence systems; it had dominated the Arab air forces in the air; and it had carried out strategic operations which were significant enough to influence Arab decisions.

The strategic attacks ranged from the bombing of airfields, radar sites and petroleum dumps to a surgical strike by Phantoms on the Syrian general headquarters, right in the heart of Damascus. The airfield strikes had an unforeseen but welcome effect: intended simply to destroy the opposing air force, in Syria they had also slowed the Soviet resupply of parts and SAMs.

The strategic war had implications which were not lost on either side, as was revealed in the Lebanese war in 1982. The bombings were extremely effective against power stations, and this in turn had adverse effects on Syria's economy for

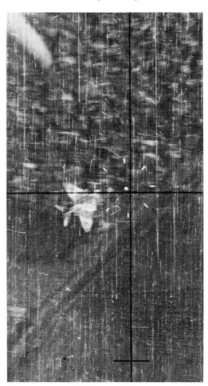
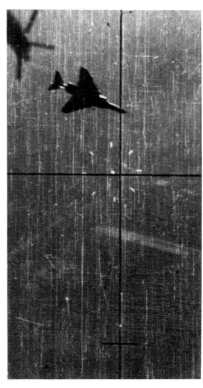

**The Israeli Phantoms suffered heavily at the hands of the Arab SA-6 batteries during the 1973 war, and the Egyptians subsequently made great propaganda capital of this tactical victory. These pieces of wreckage are on permanent display in Cairo.** *(Aviation Week and Space Technology)*

months after the war. The Syrian conclusion was to provide, in Lebanon's Beka'a area, a belt of Soviet missile and anti-aircraft systems which made all previous defences seem trivial. The Israeli solution, as we shall see, was to out-think and outfight the new missile technology.

The following article from the *Israeli Air Force Journal* of October 1981, entitled "The Hardest Mission of My Life," gives an idea of the intensity of the final days of the Yom Kippur conflict:

"It all began on October 11, when the squadron participated in an attack on Binhaha airfield in the Delta. The aircraft were caught on their way home by MiGs, attacking from their 6 o'clock. There began a vicious air battle. A few MiGs were shot down, as well as two of our Phantoms. Yonatan Ophir and Aaron Cohen, the crew of one of the aircraft, managed to eject safely but were murdered on the ground, apparently by villagers. This battle made a deep impression on the squadron. On the morning of October 14 a formation of Phantoms went to Mansura. It was clear that this was going to be difficult, perhaps very difficult. Mansura was in the heart of the Delta, defended by SA-2 and SA-3 missile batteries. MiGs were also expected in the target area and the prospects of surprise were weak.

"The first part of the approach was over the sea. As we

flew we scanned the sky constantly in an effort to see signs of MiGs or missiles. At first all was quiet, but a little before we crossed the coastline the controller called: 'Heads up. There are MiGs over your target.' 'I receive,' answered the leader. To the formation he said: 'Continuing as usual.'

"OK, now there were missiles and MiGs. The tension in each of the aircraft rose. The journey across the Delta was a long one. We flew at very low altitude, so low that from time to time you had to manoeuvre to pass over the electric cables which stretch all over the Delta. We would have to remember those cables on the way home. As we approached the target the controller reported on the MiGs in the area. No longer any question: surprise wasn't on our side. We would have to perform the attack as quickly as possible: if you waste too much time at altitude the missiles zero in on you; if they aren't able to lock on, the MiGs go for you instead. And we didn't have enough fuel to become involved.

"We arrived and broke over Mansura. We looked around. Still we couldn't spot the enemy aircraft, though we could see the MiGs' drop-tanks spinning towards the ground. We heard the MiGs above us: they could see the Phantoms and were beginning to get organised to go after the formation.

"A hard decision: whether to turn tail to the MiGs and dive for the target? During the aiming stage you fly a long time without any manoeuvres and it is very easy for an enemy fighter to set up on you. Only three days ago two of our aircraft were shot down when they were caught from behind by MiGs. Or to jettison the armament and try to escape? But then what was the purpose of this long and dangerous flight if in the end we did not attack the target?

"My No 2 and I decided to attack, hoping that the MiGs would not be able to launch until after bomb-release. We rolled in and bombed the runway. Another second and large

Above: The McDonnell Douglas F-15 has succeeded its relative of the earlier generation as Israel's front-line air-supremacy fighter. For once, the squadron insignia are not blotted out by the censor. (MDC)

Above right: The Egyptians now have F-4Es, raising the faint possibility of a Phantom-versus-Phantom conflict. (Van Geffen)

smoke columns rose from exactly where we wanted them. Stage A completed, now for Stage B, safe return to base

"We performed a hard break. There still wasn't any sign of MiGs, but we knew where they were – behind us. And then No 4 reported that he was burning. One of the pilots said afterwards in the debriefing that the first thought that came into his head upon hearing this was: 'What do you want me to do? Bring a fire extinguisher?'

"We turned towards No 4. The Phantom didn't appear to be burning, but now there were two MiGs closing the range behind him. 'No 4 break hard left. They're on you.' No 4 broke left and I broke to get on to the threatening MiGs. It's difficult when you are alone and at a definite disadvantage. That is why we talk so much about mutual defence.

"Why did No 4 think that he was burning? The answer is rather funny. His navigator was a very young man, little more than a child. He had had very few flights in the Phantom when the war started. During the attack he had heard the radio warnings of MiGs firing. Then, looking outside, he saw the condensation coming from the wings because of the high speed. He thought that this was smoke and reported it to the pilot. Hence the 'I am burning' report.

"I broke towards the MiGs and dropped the remaining armament to lighten the aircraft. I still wasn't thinking in terms of a dogfight, but only to rescue my wingman, who was in trouble. The MiGs gave up, turned towards us and then broke away. No 4 headed home, still thinking that he was burning.

"The MiG break didn't seem right to me. It appeared as if they had received a warning from someone. There were MiGs still in the area, and sure enough in another moment Baram, my navigator, warned: 'Break, they're on us.' I broke and looked behind: two MiGs, range 600-700m. Suddenly two smoke trails appeared, coming towards us from the MiGs. Air-to-air missiles!

"I moved the stick a little more. The missiles missed and the MiGs passed in front. At this point I had to decide whether to fight this pair or go home. Then Baram warned me again to break. Another pair of MiGs was on us. I didn't know if this was the first pair or if I was fighting six. I had performed two hard breaks and my speed was low. I began to worry. Another break and my speed would fall even more.

I didn't have enough fuel for a dogfight. 'No 3 stuck with MiGs,' I reported. I hoped that someone would come to rescue me. We were at low altitude and I could see the MiGs beginning to set up. I manoeuvred sharply. 'They still behind?' I asked Baram. 'Affirmative,' he replied. 'You sure?' I asked him. 'Yes,' he said, 'and now they are firing.'

"Two things were especially difficult in this battle. The first was that I was alone over Mansura, stuck with four or six MiGs, and nobody was coming to help me. Nobody. The rest of my formation had already managed to leave the area. When you're alone you sink or swim. The second thing was that the lessons of Binhaha were still clearly in my mind. This was definitely the time to leave. I had good speed stabilised in a turn, but nothing else was right. So I decided to try something else. Then Baram warned: 'Break, they're firing.' I pulled the nose up sharply and looked behind to see what was happening.

"The MiGs raised their noses after me. If they reached us we were finished. We were now completely vertical, losing speed. The MiGs were behind, but they weren't overtaking and they gave up. I raised my head: above were two MiGs waiting for me to fall off so that they could set up on us. I had to fall off because my speed was gone. This is what we call jumping from the trap into the snare.

"S, who was in his time the best pilot in the Israeli Air Force, once said: 'When I'm at a total loss I perform such a manoeuvre that even I wouldn't be able to shoot myself down if I were sitting behind me.' This was what I was going to do now. I pushed the nose down into a vertical dive. Nobody can fire in such a situation, nose-down. As the altitude fell I reached the point at which if I didn't pull up now I would never pull up. I waited a second and pulled on the stick with all my strength.

"We came out of the dive at tree-top height. The MiGs were still behind, range 1,000m. They were beginning to set up. Should I head home in military power? My brain was working at top speed. My fuel load was now less than the minimum required to return home safely. If I went to military power close to the ground they would find it easy to aim and hit us. But how far away were the electric cables? I would have to manoeuvre over them and this would enable the MiGs to get us. There was no choice: I had to turn and pass in front of them.

"I turned hard. The two MiGs were still behind and Baram warned me from time to time to break when they launched missiles or fired their cannon. Then I reverted to the manoeuvre I had performed before. I suddenly raised the nose 90°, as sharply as possible. Would they reach us this time? Fortunately they didn't, and I saw them diving away in front of us.

"We were now very low on fuel. During the last manoeuvre I had decided that I would continue to evade until I had half my present fuel load. If I hadn't disengaged by then I would head north and try at least to eject over the sea. But I now had doubts whether I would make it home on even the existing fuel.

"We continued to climb vertically. The MiGs behind had fallen away and I could see no-one in a threatening position. I lowered the nose vertically once more, and then pulled up and levelled out over the trees. I couldn't see anyone behind, and so I decided to go home, flying as low as possible. Baram watched our tail all the time.

"I want to say a little about this man of blessed memory. In this battle I saw him as he really was. Before I thought he was an ordinary navigator, but in this fight he was really exceptional. He managed to see the entire battle picture, to see the pair that I wasn't busy with and to advise me of the right thing to do. Sometimes I did what he told me and afterward I saw the MiGs. I would say to myself: 'The man was right again.' It is not easy for a pilot to admit this, but if Baram hadn't been with me I wouldn't have been able to leave Mansura safely. He simply gave me the best help that I could have asked for.

"We were now approaching the electric cables. I manoeuvred for a moment and descended back to low level. Our fuel was dwindling fast, so I cancelled afterburner and speed fell a little. The engines were now in military power, and our biggest problem was fuel. We crossed the shoreline with the fuel gauges reading minimum approaching zero. It was still impossible to go directly home because of Port Said and its missile batteries. And, sure enough, as we passed directly north of Port Said they launched a few missiles at us, but they fell into the sea.

"Where to go from here? Even Refidim was rather far. Just north of Port Said, and we had only 1,000lb of fuel left. There was only one other place, Baluza. It had one short narrow runway and was inside the missile threat area. I called: 'No 3 landing at Baluza. Prepare the runway for landing.' I silently hoped that we would find Baluza before the fuel ran out. But then No 4 suddenly came on the air: 'I am also going to land at Baluza.'

"Baluza can accept only one aircraft at a time, however, and I had only 700lb of fuel and couldn't go anywhere else. It was Baluza or eject. I reported this to No 4. 'OK,' he said to me, 'I don't have much fuel either. The first to reach Baluza can land.' I told him firmly that he would go to Refidim. He managed to land there, I wouldn't have.

"I called the Baluza control tower. Someone answered faintly: 'We hear you, where are you coming from?' It was clear that he was not used to dealing with aircraft, and I could only just hear him. Baram and I searched for the runway and started on finals. There were people and vehicles near the runway. Then the tower called: 'Go around, it's impossible to land.' I was not sure if I had enough fuel to go around but felt I had no choice. I added power for five seconds and went around. 'What's the problem?' I asked him, thinking to myself that at any moment we would have to eject. 'It's impossible to land safely,' said the man in the tower, 'there's a direct crosswind!'

"This was just too much for me: 'Idiot!,' I said, 'because of this you make me go around? You only have to tell me whether the runway is clear to land.' 'The runway is clear,' he said to me, 'but the crosswind . . .' My fuel load was now less than 300lb and from time to time the gauge would hit zero. I started on finals again, expecting the engines to cut out at any time. Another five seconds, four, three – we were down and rolling along the runway. It was like a narrow road. As

*Top:* An Iranian F-4 on its delivery flight across the Atlantic. When Capt McGregor Poll landed at Tehran he noticed a 727 standing in the parking area, with the Shah on the rear airstairs. By the time Poll had unbuckled and climbed out of the aircraft, the American symbols had been washed off and the Iranian flag painted on. *(Dr McGregor Poll)*

*Above:* Poll flew the Atlantic to Iran with back-seater Capt Lanny Lengwith. Having picked up the aircraft at McDonnell Douglas St Louis, they then flew to MacDill AFB, Florida, to be briefed on the flight to Iran. The aircraft took off at one-minute intervals and flew in trail. They tanked up over Bermuda and the Azores and landed at Torrejon, Spain, after 8.5hr flying time. A few days later they flew the last 5.5hr to Tehran. *(Dr McGregor Poll)*

*Right:* Some of the most highly capable American aircraft went to Iran. Here an F-4D taxis in at Mehrabad AFB, Tehran. Iranian ground crews were originally trained and supplemented by US technicians, so that even the most complex aircraft were at a high state of readiness. *(John Harty)*

we turned off the runway the engines cut. We had done it! We left our cockpits and embraced."

After the Yom Kippur War both the Israelis and the Arabs went through some agonising reappraisals. The Egyptians wanted to know how their original winning strategy had gone wrong. The Arabs had learned much from the War of Attrition, and had formulated a plan based on a massive infusion of Soviet weapons of the finest quality. The SAM defences were immeasurably improved by electronic counter-countermeasures and tactics that seriously reduced the previous Israeli superiority in the field. Then the SA-2s and SA-3s were integrated with the modern SA-6s and SA-7s, plus the even more effective Shilka quad-mounted radar-directed mobile 23mm anti-aircraft vehicles. These deadly weapons, along with SA-7s, were deployed heavily in the valleys known to be favoured by the Israelis for their low-level approaches to the attack.

A large passive defensive effort was also made. Command posts, power sources and radar display areas were covered with heavy concrete structures, then camouflaged with sand. The defensive systems were very mobile, and were combined with an extensive network of dummy installations which made Israeli targeting difficult.

Finally, and most important, the Arab forces were trained to unprecedented heights of efficiency. All of this was designed to stop in its tracks an Israeli pre-emptive attack. This did not materialise, however, with the result that a fatal flaw in the Arab strategy was exposed, particularly on the Canal front. The network of defences, formidable as it was, could not be advanced as a unit without losing a large measure of its effectiveness. As the Egyptian forces pushed forward in an effort to relieve the pressure on Syria, the integrity of the defensive network was broken and it was chewed to pieces by the Israelis.

A secondary problem for the Arabs was their inability to resupply their batteries with enough missiles and ammunition to sustain the profligate rate of fire indulged in during the first three days. After the IAF's initial battering Gen Peled ordered the squadrons to desist from attacks on the Canal line until new tactics could be developed. This effort included a ground breakthrough by armoured units on the west bank of the Canal that overran SAM and flak batteries and provided "elbow room" for further air attacks.

Most observers drew from the Yom Kippur War conclusions that were pessimistic about the future of Israel. She had prevailed once again, thanks to her air force and the rapid recovery by her ground forces. But the margin had been terribly narrow. The Arab forces were obviously significantly better than ever before in terms of training, morale and equipment. They had failed once again to establish the necessary co-ordination, and their conduct of the battle had shown that the individual Israeli soldier and commander were still better able to execute independent operations. But the difference between the Six-Day War and the Yom Kippur conflagration was obvious. The Arabs were closing the qualitative gap, particularly in missile defences, while their quantitative advantage would continue to increase on account of the new-found oil wealth of the Arab countries, which for almost a decade permitted them to buy whatever was available, and also because they were committed to using less expensive Soviet equipment. A MiG-21 costs about $2 million (or even as little as $800,000 on the black market, it is said), compared with about $5.7 million for an F-4; a MiG-23 comes out at about $7 million compared with $30 million for an F-15. Israel can never close the quantitative gap, so to survive she must maintain a qualitative edge, a task made all the harder by the introduction of F-4s into the Egyptian Air Force and F-15s into the Royal Saudi Arabian Air Force. When Awacs, F-5Gs and F-16s are added to the equation, the outcome of any future Middle East war becomes very hard to predict.

Events in the Lebanon in 1981 and 1982 were to prove that the Israeli Air Force would continue to "find a way" in the face of the ever-lengthening odds. Primarily responsible for the Beka'a triumph was Colonel A, who also planned the attack on the Iraqi nuclear reactor.

The Syrian defences were formidable, an adroit mixture of 19 SAM batteries and many 23, 37 and 57mm flak guns. The most dangerous weapons were again the SA-6 Gainful and the ZSU-23-4 quad-mounted anti-aircraft gun. But

Left: F-4D with 20mm centreline gun pod. The Iranians bought 32 F-4Ds, 177 F-4Es and 23 RF-4Es, and possessed the most formidable striking force in the Middle East. But when the Shah was deposed its efficiency declined precipitously. *(John Harty)*

Below: An Iranian F-4E. At first these aircraft were meticulously maintained. *(John Harty)*

while the batteries were in most cases mobile, resulting in some desirable defensive attributes, they were no longer protected by heavy concrete bunkers and emplacements, making the Israeli task easier. Further, the Syrian Air Force decided to protect the batteries with large forces of MiG-21s and MiG-23s. But if the emplaced missile batteries were intended to destroy the oncoming Israeli Air Force, why did they need to be protected? And if they needed to be protected, why were they acquired at all? For the emphasis on the SAM batteries, with their associated radar and command and control network, had imposed a severe personnel burden on Syria; trained men are in short supply, and they were diverted wholesale to the Beka'a defences.

The air battle proved to be a slaughter, with reports of as many as 100 Syrian aircraft being destroyed without any losses to the Israelis. IAF pilots commented that the Syrian pilots appeared to be even less well trained in the use of their aircraft than on previous occasions. Israeli E-2C Hawkeyes were a vital factor in the combat, enabling the fighters always to be on the spot at the right time.

The destruction was just as great on the ground. Israeli tactics, based on the use of as many as 150 attacking aircraft in combination with ground forces, laid waste to the missile system. The Israelis used cluster bombs, TV-guided Maverick missiles and greatly updated electronic counter-measures equipment, all applied with a precision that reflected years of combat experience and hundreds of practice missions.

The F-4 proved itself again to be the master of the flak batteries, being the principal weapon in the attacks. The aircraft themselves, their Shrike and Sidewinder missiles, and their ECM equipment had all been subjected to Israeli-developed improvements; but the key to success lay principally in a decision by the IAF command. The most important difference between the Beka'a victory and previous F-4/missile battery contests resided in the instructions received by the pilots before the battle. Air combat was to take a back seat until the missile defences had been destroyed. As a result, 19 batteries of SA-2s, 3s and 6s were knocked out by a massive June 9 strike, most within the first two hours. Subsequently the IAF was able to deal almost at leisure with the attacking MiGs, while artillery fire was added to the weight of aircraft attacks against the optically operated SA-6s. To sum up, the attack on the Beka'a succeeded because the IAF drew the correct conclusions from the bitter experience of the Yom Kippur War.

The Soviets were greatly embarrassed by the collapse of the missile systems and fighter forces that they had set up. The immediate reaction was of course to blame the Syrians for lack of heart, and then to begin a massive resupply effort. It was all in vain, though, for on July 24, 1982, the Syrians experienced the same sort of joy at the discomfiture of the Soviet forces that their Egyptian comrades had felt in July 1973, when Soviet-piloted MiG-21s had been trounced by Israeli Mirages and F-4s.

The Russians rushed in SA-8 batteries to replace the Syrian losses in SA-6s. Three SA-8 batteries were installed in the Beka'a and were almost certainly operated by Soviet crews and technicians, the Syrians not having had adequate experience on the type. IAF F-4s attacked the batteries and destroyed them, just as they had the previous installations, leaving the Russians without either weapons or excuses.

Naturally, the battle against the Beka'a missile and AAA defences has attracted most attention and analysis. But the following air-combat account, from the *Israeli Air Force Journal* of February 1983, is of great interest, particularly because it depicts the Phantom jostling for a kill with its mighty McDonnell Douglas descendant, the Eagle: "Friday, June 10, 1982, the last day of intense air activity during Operation Peace for Galilee, late morning. Two Phantoms were on a routine reconnaissance in Lebanese skies. In the west the artillery still thundered and thick smog covered Beirut, a reminder of the massive shelling of the previous night. In the east there was nothing new. Here and there smoke columns rose in the area of Lake Karoun.

"The Phantom pilots didn't expect any surprises. The great hours were already behind them. One of them was Lt-Col P, a veteran Phantom pilot who already had four kills behind him. P had learned never to expect too much, though he was keenly aware that this was perhaps the last chance for the Phantom to shoot down something in this war.

"'There wasn't anything for us to be ashamed of,' he says. 'On the contrary, the Phantoms distinguished themselves in the attack missions during the beginning of the war. But we wanted meat and this patrol was the sort of thing that was thrown to us.'

"The tension which is characteristic of the take-off faded away as the Phantoms flew on tranquilly, accepting the directions of the controller. But then the tranquillity evaporated as the controller suddenly turned them east towards unidentified aircraft.

"'We dropped our tanks, went to afterburner, and proceeded east at full power,' Lt-Col P continues. 'From my experiences in the past I knew not to raise my hopes too much. But my excitement grew from the beginning of the "warm-up" and continued to rise as speed increased. As we headed towards the unknown, eyes scanning the horizon for prey, our thoughts were concentrated on the mission, everything else forgotten. All my attention was directed at spotting the target.'

"But then came disappointment as the Phantom crews failed to spot the targets. The Phantoms had turned for home when the controller called them again, directing them to a target in the west, at low altitude. 'We went to afterburners again, and flew west until we saw them, two brown dots over Lake Karoun, heading towards us. Later we found that these were two MiG-21s that had tried to intercept our helicopters.'

"But D's formation wasn't the only one turned towards the MiGs: two F-15s were also racing in to intercept. It was Phantoms against Eagles. 'The "F" beat me to a firing position on one of the MiGs, launched first and missed. I launched immediately after him and missed also. The MiG was now very close to me, but it was the F that launched last and the MiG exploded into debris and smoke in front of me. More disappointment.

"'The second MiG had disappeared, apparently heading west at low altitude. The controller turned us towards him. We turned hard and the Fs, because of their superior performance, left us behind. There was nothing for us to do but search for the MiG on the radarscope. We couldn't see anything and I was worried that our last chance was slipping away. I told the navigator that I was giving up on the radar and trying to acquire him visually. It was now or never.

"'After some seconds I saw him: a dot over Jebel Bruk, heading towards us. We weren't sure if this was a MiG until it launched a missile at the F-15s, who saw it and broke. The missile passed them and hit the ground. Now the MiG was coming directly towards me. Was he aware of me? But all his thoughts were on the Fs and he passed in front of me, heading south to north, trying to evade the Fs and not thinking about anything else except his success in fooling the two Eagles.

"'I broke hard right and after a short period I was sitting on his tail. From here on everything was simple. I pursued him until he was centred in my sights, and then I launched. I waited long seconds as the missile headed off on its way to the target and I tried to guide it by willpower. I waited for the MiG to break and the missile to fly on harmlessly. But this didn't happen: the missile entered his engine and blew the MiG to pieces.

"'End of story. On the way home I reported that I had had a kill and was coming in for a buzz (victory roll) over the base. They had not yet heard about the kill. We quickly organised a celebration'."

The implication of the Beka'a victory were perhaps overshadowed by the subsequent draining events in the Lebanon. Still, the Israelis drew some conclusions that could profitably be applied around the world. The first of these is that the aircraft, with adequate electronic warfare support, still dominates, for it is infinitely more mobile than the most mechanised SAM system and, as the IAF illustrated, can operate with other types in a tactical harmony that overcomes the disadvantages of ground-based radars and ECM equipment. The second conclusion is that there is no substitute for arduous, realistic training, despite the fact that it is expensive in material and often in men.

The Phantom has now proved itself in more than a decade of service with the IAF. Though it has now yielded the limelight to the Eagle and F-16, it is still Israel's leading combat type in actual warfare. The Israelis value the F-4 as the aircraft which transformed the IAF from a defensive arm to one capable of exerting pressure not just on Suez and the Golan but also on Cairo and Damascus. The F-15s and F-16s may be more glamorous, but the down and dirty fighting will be carried out in large part by Phantoms for years to come.

The successful employment of the Phantom in the Vietnam War and the Israeli/Arab conflicts depended on discipline, training and mammoth supply networks. Though the type's fortunes in the protracted, bitter and, to Western eyes, senseless Iran/Iraq conflict have been less well documented, it is almost certain to have performed nowhere near as well as it did in Israeli or American hands. The Iranian Islamic Revolutionary Air Force is but a shadow of the Shah's Imperial Iranian Air Force. Even so, the Phantom has persisted as the most formidable weapon in the Iranian air arsenal for a far longer time than had been projected. While being unable to maintain its 177 F-4Es, 32 F-4Ds and 16 RF-4Es to anything like Israeli standards, the reduced IIRAF has nevertheless been able to sustain a limited offensive with the Phantom since the war began. The general consensus is that the Iranian F-4s fly and fight with far less reliable equipment and arms than one would find in either Israeli or US units, but are nonetheless able to penetrate Iraqi airspace almost at will.

In the days of the Shah the Imperial Iranian Air Force worked hand in glove with its American suppliers to create what appeared to be one of the most formidable striking forces in the Middle East. The 200-plus F-4s were supplemented by 80 Grumman F-14A Tomcats, with their associated Phoenix and Sidewinder missiles, and more than 150 Northrop F-5s. These combat types were supported by Boeing 707-320L tankers, Lockheed C-130 transports, helicopters and maritime reconnaissance Lockheed P-3C Orions. The Shah had intended to become the superpower of the oil-rich Golden Crescent, backing up his offensive forces with excellent surface-to-air missiles and anti-aircraft artillery. An early-warning system of ground radars was to have been further enhanced by the acquisition of Boeing E-3A Awacs aircraft. The pilots and ground crews of the IIAF were well trained, and supplemented by well paid US contractor personnel.

The Iraqi Air Force (*Al Quwwat Aljawwiya Aliraquiya*) is largely equipped with Soviet aircraft and missiles but supplements these from a variety of sources so that its inventory includes no fewer than 37 different aircraft types from several countries, including France, Britain, Switzerland and Czechoslovakia. At the start of the war the strike force comprised squadrons of Tupolev Tu-22 and Tu-16 and Ilyushin Il-28 bombers, plus 160-plus fighter-bombers. These included 75 MiG-23s, 50 Sukhoi Su-7Bs and 30 Hawker Hunters. In addition there were about 115 MiG-21 and 36 Mirage F.1 interceptors. The war has resulted in heavy losses, and replacements come largely in the form of additional MiG-23s and Sukhoi Su-20 variable-geometry fighters.

Neither side has shown the élan that one might expect of warriors in a holy war. This is probably on account of the complexity of the equipment: there has been no shortage of bravery – or butchery – on the ground. Air-to-air battles have been infrequent and inconclusive. The fighter-bombers have been used mainly for strikes against oil depots, pipelines and refineries. Both sides attacked airfields early in the war, but neither achieved the degree of success essential for even temporary air superiority.

The war began with an ill-conceived Iraqi attack on Iranian airfields which failed entirely to achieve the classic objective of destroying the enemy on the ground. The Iranian riposte was much better handled and did hurt the Iraqi air force, causing it to disperse to bases outside the country. Victory claims by both sides were absurd: had they been genuine, both air forces would have been annihilated very early in the war. The Iranian Air Force, which had suffered heavily following the fall of the Shah, received a reinfusion of talent when about 100 of its pilots were released from prison to fight. But with only 20-50 per cent of its aircraft operational to begin with, it could not possibly have achieved what the Shah had intended for it.

The Iraqi Air Force was inferior on all counts: the aircrews had obviously received inadequate training in bombing, ground attack and air combat, and its radar and communications networks were very inefficient. It was evident that, while the Soviets had supplied adequate equipment, they had not backed it up with the necessary training, discipline and spares support.

There were instances in which Iranian Phantoms distinguished themselves. They were able to strike airfields and oil stores, and once caught an Iraqi armoured formation massing for the attack. Further, they have projected a strategic shadow over both Iraq and Kuwait. Being able to reach almost anywhere in Iraq, the Phantoms have to be reckoned with even in their depleted state. A series of raids against Kuwaiti oil facilities gave a clear signal to neighbouring Arab states that Iran was not to be trifled with.

At the time of writing the war goes on, with the air forces following a desultory pattern of stand down, recuperate, and then launch a strike force of four to eight aircraft. Iran has kept its dwindling fleet fighting by buying spares in every market available, including Israel. Curiously, the main source is North Korea, which is almost certainly dealing in American stores captured in Vietnam. The Iraqi Air Force seems to be showing slightly more recuperative powers, and daily sortie totals of up to 150 have been seen. The effectiveness of the Iraqi attacks has not however risen significantly. In general, the Phantom has performed better than any other aircraft in the theatre, but at nowhere near the levels seen in other wars. It is really not a war worthy of the Phantom.

# 9. A peacetime winner too

What has given the Phantom its edge over the opposition in every war it has fought? Crew quality, tactics and weapons have all played their part, but ultimately it came down to one thing: performance. The potential available to Phantom users was immediately apparent, and the two American services decided to exploit it by setting a whole host of records. This had two purposes: every record served as further reassurance to the Navy and then the Air Force that their choice had been right, and could be used as ammunition in the campaign for more Phantoms in Congress. Similarly, foreign sales were unquestionably influenced by the spate of records.

The power-to-weight ratio of the Phantom was so good that an altitude record seemed to be the most readily attainable. So on December 6, 1959, Cdr Lawrence E. Flint Jr took off from Edwards Air Force Base on a flight which formed part of the fleet testing programme. He climbed almost vertically and levelled off at 50,000ft, where he accelerated until cleared by the ground to pull up into a ballistic trajectory. He hurtled upwards, experiencing weightlessness at the top of an arc which carried him to 98,557ft and a minimum indicated airspeed of only 45mph. The previous record, 94,568ft, had been held by the Soviet Union.

Lt-Col (later Lt-Gen) Tom Miller USMC set the first Phantom speed record on September 5, 1960, when he flew a 500km closed course over the Mojave Desert at an average speed of 1,216.8mph. Compared with the 311 measured miles (500km) of the offical course, the Phantom actually covered 334 miles air distance; time on the course was 15min 19.2sec.

The Navy set the next record, on September 25, 1960. Lt-Cdr J. F. Davis flew a 100km closed circuit at 1,390.24mph, Mach 2.21, holding a 70° bank and pulling a constant 3g all the way around. The 100km circle was completed in 2min 40.9sec.

**Above right:** The Phantom II Spook on a 1,000hr patch. The figure was designed by McDonnell Douglas artist Tony Wong in 1958, and it found immediate acceptance around the world. It has been used on aircraft shelters, playing cards and T-shirts, and the Spook has been depicted holding machine guns, cameras and fire hoses.

**Right:** Phantom II Spook on the tail of an RF-4B. The Spook originally held his finger is a somewhat ruder manner. *(Don Linn)*

Project Lana – signifying the 50th (L) (A)nniversary of (N)aval (A)viation – was a modern rerun of the old Bendix air race. Five Phantoms took off on May 24, 1961, from Ontario Field, Los Angeles; three were to be competing aircraft, the other two airborne spares. The Phantoms refuelled three times before the first of them, crewed by Cdr J. S. Lake and Lt E. A. Cowart (RIO), flashed across Floyd Bennett Field, Brooklyn, NY, after 3hr 5min, a new record. The second team across broke this brand-new record: Cdr L. S. Lamoreaux and Lt T. J. Johnson required only 2hr 50min. But even this blistering time was bettered when Lts R. F. Gordon and B. R. Young finished in just 2hr 48min for an average speed of 869mph.

Two young men with movie-star names, Lt Huntington "Hunt" Hardisty and Lt Earl H. "Duke" DeEsch, set a low-altitude speed record over a 3km course in the Stallion Sight area at White Sands Missile Range, Alamagordo, New Mexico, on August 28, 1961. The course was hazardous, and an F-4 had broken up and crashed on a previous attempt. The rules required that the aircraft stay below 100m (328ft) above the ground throughout the record run. The crew flew the aircraft twice in each direction at a maximum altitude of 125ft above the ground to set a new world record of 753.64kt (Mach 1.18).

Perhaps the riskiest of the record attempts was Operation Skyburner, in which an F-4 modified with pre-compressor cooling set a blazing record of 1,606.3mph on November 22, 1961. A stainless steel tank was filled with water and placed in the rear cockpit; the water was introduced via a spray bar to provide pre-compressor cooling, permitting the Phantom to reach an official Mach 2.5 and an unofficial Mach 2.64. The pilot was Lt-Col R. B. Robinson USMC, who had to use all of his great skill to keep the aircraft under perfect altitude and airspeed control during the record run.

Cdr George W. Ellis USN set a sustained altitude record of 66,443.8ft on December 5, 1961. Cdr Ellis was required to make one pass over the FAI course without decelerating or descending more than 100m from entry to exit. He was cruising at 40,000ft 180 miles from the course when he lit both burners and accelerated until crossing the gate. While still 40 miles from the entry gate the aircraft was at 60,000ft and had achieved Mach 2.2; the aircraft crossed the gate at 66,443.8ft and exited at 66,237.8ft, breaking the previous record by more than 11,000ft.

From February to April 1962 the Navy and Marine Corps set a whole host of time-to-height records in Operation High Jump. They ranged from Lt-Cdr John W. Young's 3,000m record of 34.52sec, set on February 21, to Lt-Cdr Del W. Nordberg's 30,000m mark of 371.43sec, set on April 12, 1962. (Young is now Nasa's senior astronaut, having flown in space six times at the time of writing.)

Another record was set in 1979, when an RAF Phantom flew the Atlantic from Goose Bay, Labrador, to Greenham Common, England, in 5hr 40min, emulating the 1919 flight by Alcock and Brown. The crew on this flight was, appropriately, Sqn Ldr Tony Alcock and Flt Lt Norman Browne.

All this capability didn't come cheap, however, and the Phantom was an expensive fighter from the start. But it soon became apparent that the unit cost might be lowered if sales were made to friendly foreign governments. Sometimes these sales also made it possible for production lines to operate at the optimum rate, so keeping the workforce constant and supporting subcontractors.

Thus McDonnell was more than happy to initiate a new marketing, maintenance and supply programme, beginning in early 1964 with the sale of 52 F-4Ks and 118 F-4Ms to the United Kingdom. These aircraft were the most modified of all the Phantoms, incorporating the Rolls-Royce Spey engine and an airframe of which nearly half was produced in Britain.

The Imperial Iranian Air Force became the third largest user of Phantoms after the United States and Germany, buying 32 F-4Ds for delivery between May 1968 and

October 1969 and 177 F-4Es between April 1971 and August 1977. The Shah also bought 23 RF-4Es between December 1971 and October 1979. While they were operated by the Imperial Iranian Air Force, with suitable American contractor support, these F-4s were a formidable force. After the Islamic revolution many members of the officer corps were imprisoned, while contractor personnel were either imprisoned or expelled. As a result, aircraft availability plummeted. Then came the war with Iraq: many of the detainees were released back to active service, and F-4 availability went up. After years of war, however, it is estimated that perhaps only 20 per cent of the force is truly operational.

The Republic of Korea has operated 36 F-4Ds since 1969, and 37 F-4Es since 1979. Under US tutelage the RoKAF has

Below left: The Phantom set a host of records. Among the first was an absolute altitude mark of 98,557ft set on December 6, 1959, by Cdr Lawrence E. Flint Jr, flying from Edwards AFB in the course of Operation Top Flight. Cdr Flint levelled off at 50,000ft, accelerated and, on command from the ground, zoom-climbed into a ballistic trajectory that carried him to the record. *(MDC)*

Below: Project High Jump was conducted in 1962 from Brunswick and Point Mugu naval air stations. Several different Phantoms captured eight time-to-height records, including 30,000m in just 371.43sec, on April 12, 1962. *(US Navy)*

Right: The record set by Duke de Esch and Hunt Hardisty (the improbable but genuine names of the lieutenants who flew Sageburner) – 902.769mph over 3km at low level – lasted for many years until Darryl Greenamyer beat it in his "homebuilt" F-104. *(MDC)*

Below right: Hardisty and De Esch beside their record-setting Phantom. *(MDC)*

Above: McDonnell probed the outer limits of the Phantom's structural strength with Skyburner, which achieved 1,606.3mph and, reputedly, Mach 2.64 on November 22, 1961. Lt-Col R. B. Robinson USMC set the record at Edwards, using a top-secret system installed in the rear crew position. Cooling water was sprayed into the engines' pre-compressor sections, just as in the RF-4X. *(MDC)*

Below: In a brilliant piece of public-relations opportunism the RAF had "Alcock and Browne" repeat the pioneering transatlantic flight 60 years later in a Rolls-Royce Spey-powered F-4. *(MDC)*

Below: McDonnell celebrated each major production milestone with a mixture of joy and incredulity. Who would have believed the 1,000th Phantom in 1958, much less the 5,000th in 1978 and the 5,057th and last the year later? *(MDC)*

Above: McDonnell and the Navy also celebrated the first flights of new models, as these pictures of the F-4J and F-4N show.

achieved a high degree of proficiency. The Korean Phantoms have not yet gone into action; there have been some tentative engagements with North Korean fighters, but no extensive conflict.

Israel, as we have seen, has operated the Phantom more effectively than perhaps any other user nation, thanks to intensive training and a history of almost unremitting conflict.

The Royal Australian Air Force, left short of capability by problems with its prospective F-111 buy, quietly leased a stopgap force of 24 F-4Es from the USAF, beginning in September 1970. These aircraft have since been returned to the USAF.

The Luftwaffe, badly shaken by its experience with the crash-prone F-104, turned to the twin-engined Phantom with no little relief. The Germans bought 175 F-4Fs between May 1973 and February 1976. A result of the US International Fighter Competition of 1962, the F-4F is essentially a version of the F-4E with simplified systems and armament. The Luftwaffe also bought 88 RF-4Es in 1970 and 1971 and 10 standard F-4Es in 1977.

The Japanese were as much interested in acquiring manufacturing techniques as they were in the aircraft itself, and arranged with McDonnell and the US Government for two F-4EJs to be built entirely at St Louis, eight more readied for assembly in Japan, and the remaining 130 produced there.

Other air forces followed rapidly: the Greek Air Force acquired 56 F-4Es and eight RF-4Es; the Spanish Air Force received 36 F-4Cs and four RF-4Cs, to become the only country besides the US to operate the C model. The Turkish Air Force operates 87 F-4Es and eight RF-4Es. And, in an odd turn of the Middle Eastern wheel, Egypt now has 35 F-4Es.

The Phantom is a good example of what happens when an aircraft turns out to be superlative: people want it to be even better. The McDonnell Douglas official listing of proposed model numbers for the F-4 covers almost 400 variants, from 98A, the proposed F3H-E, through 98 DB, an Army close-support F4H, and 98 FJ, a "very advanced F-4B" for the Navy with Pratt & Whitney TF30 engines, to 98 PF, an advanced Air Force RF-4.

As early as 1965 McDonnell made presentations of swing-wing and V/Stol developments which ranged from the conventional to the wild. Much more serious perhaps were the F-4 (VS) and F-4M (VS) proposals, which offered slightly better performance than the conventional F-4 for an acceptable increase in cost. But the demand for F-4s continued at such a level, and the capability of the basic F-4 was so great, that no go-ahead was given.

The less radical RF-4X offered enormous performance increases through the use of pre-compressor cooling, as applied to the record-breaking Operation Skyburner aircraft. The RF-4X incorporated an improved version of the Skyburner cooling system, as well as the very advanced HIAC camera. The RF-4X was intended to have a top speed of better than Mach 3 and a steady cruise of Mach 2.4 for at least 10min. Some 5,000lb of water was to be carried in a saddle tank over the spine of the aircraft.

Curiously, the F-4 was the victim of its own superiority. It was needed so badly and in such quantity that many basic improvements, such as the elimination from USAF aircraft of Navy equipment (folding wing, arrester hook, etc) were not attempted simply because it was easier to keep them. During its long production lifetime, no opposing fighter really came to equal the Phantom, and the changes that were introduced (slats and new engines, armament and equipment) were sufficient to keep the basic design competitive. Nevertheless, the McDonnell Douglas engineers were well aware that much could have been done to improve the aircraft. As it happened, many of these features were incorporated in the follow-on F-15 and F-18.

Asked what they might have done differently on the F-4 if time and money had permitted, veteran McDonnell engineers came up with the following:

1 Make the aircraft four feet longer to accommodate more equipment.
2 Use solid-state devices instead of vacuum tubes.
3 Use digital instead of analogue computers.
4 Use a multi-mode instead of a non-coherent-pulse radar.
5 Fit a head-up display instead of the fixed gunsight.
6 Use an autopilot with redundant channels and stability and control augmentation.
7 Use TV-type cathode-ray tubes for radar and other displays.
8 Use a multiple bussing system instead of individual electrical wires.
9 Use electronic instead of electro-mechanical flight displays.

Many of these improvements are incorporated in the F-15 and F-18, which also benefit from materials and techniques pioneered on the Phantom. Experience with the F-4 has insured that the Vietnam potting compound problem won't be repeated. Similarly, the lessons learned while fabricating titanium for the F-4 have resulted in vastly improved techniques for the Eagle and the Hornet. The chemical milling methods worked out for the F-4 proved fundamental to the construction techniques used on the F-15. The F-4 tested both beryllium and the boron epoxy composite rudders, the latter being preferred because of their lack of toxicity. Fifty rudders were installed on USAF F-4s for tests, and the results were so convincing that the F-15 is fitted with composite tail surfaces as standard.

The F-4 was the first McDonnell product to make extensive use of large integral machine parts instead of assemblies of many bits and pieces. This resulted in a substantial weight saving and, even more important, faster assembly and more resistance to corrosion and fatigue. As a single example, the inner wing spar would have consisted of at least 45 smaller parts held together by over 700 fasteners. The F-4 instead incorporated a forging which was bent to a dogleg shape to suit the wing planform, over 16ft in length, and machined all over. This was stiffer and more fatigue-resistant, 121lb lighter and significantly cheaper.

**Below: An F-4S of VF-302, a reserve squadron based at NAS Miramar.** (Don Linn)

**Above:** A Spey-powered Phantom launches from *Ark Royal* in Royal Navy colours. *(MDC)*

**Above right:** Relatively few fighter squadrons have flown the same type for 20 years. The US Navy's VF-74 is proud to have put in two decades on Phantoms. Times have certainly changed: can you imagine the 94th Aero Squadron still flying Spads in 1938? *(Don Linn)*

**Right:** An F-4K with fintip-mounted ECM gear. *(Harry Gann)*

**Below:** Turkish F-4Es pictured in September 1981. *(MDC)*

Top: The Egyptians far prefer to see their own F-4Es flying over the Pyramids. *(MDC)*

Above: Luftwaffe F-4Es in formation over Goose Bay, Labrador, during a Nato exercise in September 1982. The Luftwaffe flew CAP for USAF C-141B and C-130E transports. *(Don Linn)*

Above: An RF-4EJ of the Japan Air Self-Defence Force. *(MDC)*

The Phantom structure also contained aluminium forgings that required no direct machining and replaced many machine parts. The F-15 and F-18 designs benefited directly from this experience: there were over 300 such "no-draft" precision forgings in each aircraft. More important from a national point of view, no fewer than 12 companies were established to provide similar parts for other manufacturers.

The McDonnell engineers extended the capabilities of aluminium forgings by introducing stress relief by means of a compression process that stabilised the material and made it more easily controlled during machining.

There were hundreds of other similar advances, including better integral wing fuel tank seals, improved casting techniques, better paints, and so on. But perhaps the most important legacy was the insistence that the Eagle and Hornet be immune to departure, the stall/spin phenomenon that had plagued operation of the Phantom at high angles of attack. For these new aircraft the problem simply does not exist, thanks to better cockpit and control design, and superior black boxes which "inform" the aircraft of what is happening and initiate corrections long before a human pilot could possibly react. Other valuable Phantom bequests include better electronics, smaller radar and infra-red signatures, no smoke, and far more survivability. The F-4 had probably contributed more to US fighter design than any previous type: its strengths were built upon and its weaknesses eliminated.

The Phantom was for a long time the premier jet fighter in the world. During that time it enjoyed a margin of superiority that was probably greater than that of the Sabre after the MiG-15, the Zero over the Brewster Buffalo, and the "Fokker Scourge" over the Allied fighters on the Western Front. For an even longer time it has remained a top contender, able to compete effectively against even the most modern fighters under certain conditions. It began as a speculative effort by McDonnell to stay in the fighter business; production ended after 5,195 had been built and delivered to ten different countries. The type is still being improved in massive modification programmes and re-issued under new designations. It has passed to US Air National Guard and Reserve units, but it also soldiers on in first-line USAF and Navy service. Around the world, the Phantom is still an aircraft to be reckoned with. Few aircraft have done so much, under such varied conditions. Even today, few types can face such a certain future. The year 2000 will come, and Phantoms will still be doing their job.

But that is not all: the Phantom brought about the greatest revolution in air combat in history. It offered so much potential, demanded so much of its pilots, that special training units had to be set up to extract from the aircraft the maximum possible performance. These units – Red Flag for the USAF, at Nellis AFB, Nevada, and Top Gun for the US Navy, at NAS Miramar, California – have produced fighter pilots who have no equals in the world. It was the Phantom's huge latent capability, its hitherto unexploited excellence, that brought these air-combat universities into being. "Sometimes too good for even the best": will that be history's verdict on the hulking prodigy from St Louis?

Below: Australia leased Phantoms from the US when delivery of its F-111s was delayed. *(MDC)*

**Top:** By famous aviation artist Keith Ferris, this paint scheme is designed to confuse in combat. *(MDC)*

**Above:** Phantoms have been flown with great dash by both the US Navy's Blue Angels (shown here) and the USAF's Thunderbirds. *(MDC)*

**Below:** During the US bicentennial year Phantoms were decorated in coats of many colours. *(MDC)*

Below: In-flight photos of the still-classified **F-4G Wild Weasel** are rare. Constant improvements are made to the type's electronics to keep pace with Soviet developments. This aircraft is carrying no fewer than four different types of anti-radiation missiles (from starboard outboard pylon: Shrike, Standard ARM, Maverick and HARM). *(via Al Lloyd)*

Left: The most modified Phantom was undoubtedly 62-12200. Originally intended to be a Navy F-4B, it became the prototype USAF RF-4C. After completion of tests as the YRF-4C it was transformed into the F-4E prototype, the cameras being removed and the centreline cannon added. Later it tested the air-combat manoeuvring slats in Project Agile Eagle, before being fitted with "fly-by-wire" electrically signalled controls. Finally it sprouted canard control surfaces for the Precision Aircraft Control Technology (PACT) programme. On January 9, 1979, it was ferried, dangling beneath a US Army CH-54B helicopter, to the Air Force Museum at Wright-Patterson Air Force Base, Ohio. *(MDC)*

Right: The swing-wing potential of the F-4 captured the imagination of McDonnell and the US services for a long time. But the advantages never quite outweighed the need for uninterrupted F-4 production. *(MDC)*

Below: RF-4X testbed being prepared for investigations of precompressor cooling. The aircraft was a highly modified Israeli F-4E, 69-7576. A top speed of Mach 3.2 was expected. *(Jay Miller)*

Bottom: Many of the results of development work on the F-4 found their way into the F-18 Hornet. *(Harry Gann)*

**Above:** Owing much to the Phantom, the F-15 had its first taste of combat with the Israeli Air Force and quickly established its superiority in the Middle East. These gun-camera shots reflect an F-15A/B gun and missile engagement with Syrian MiG-21MFs on June 27, 1979. Lt-Col Eitan and General Eliyahu scored a gun kill against a MiG-21. *(MDC)*

**Left:** The starboard intake ramp of 66-338 boasts of 18 F-15 Eagles bagged in simulated air combat. *(Fred Johnsen)*

Left: The USAF gives its own and allied aircrews realistic combat training in the Red Flag exercises held at Nellis AFB, Nevada, every year. F-4E 66-338 of the 18th TFS carries a variety of "kill" symbols, including a B-52 at top right. *(Fred Johnsen)*

Top: A Top Gun F-5A "escaping" from VMFA-323 "Death Rattler" F-4Ns. *(Harry Gann)*

Above: The Phantom soldiers on: Oregon Air National Guard F-4s photographed on May 16, 1982. *(Fred Johnsen)*

# Appendix 1
# Phantom variants

| Model | Quantity | Notes |
|---|---|---|
| **XF4H** | 2 | First flight May 27, 1958. Bob Little pilot. Two General Electric J79-GE-3A engines on loan from USAF, 9,300lb of thrust. First flight from carrier USS *Independence* Feb 15, 1960, Lt-Cdr Paul Spencer. Second article (BuNo 142260) lost on altitude flight, killing Zeke Huelsbeck. |
| **F4H-1F** | 5 | Pre-production standard. BuNo 143392 first to have boundary-layer control installed. Redesignated F-4A. J79-GE-2 or 2A engines, 10,350lb thrust. (Second "F" designation indicates "special engine".) |
| **F4H-1F** (F-4A) | 40 | Production aircraft. First 11 shared with initial 7 the low-drag canopy, APQ-50 radar with 24in antenna dish. Aircraft Nos 19-47 had raised canopy for better visibility, 32in-dish APQ-72 radar. All redesignated F-4A on Sept 18, 1962. First 26 of F-4A series used as R&D aircraft; remaining 21 issued to squadrons for training. Six Sparrow III air-to-air missiles. First user unit was VF-101, December 1960. Some ultimately redesignated TF-4A. |
| **F-4B** | 649 | Received J79-GE-8 engines, 10,900lb thrust without afterburning, 17,000lb with. Four Sidewinders to complement Sparrows. 16,000lb of ordnance. 12 converted to F-4G (see below). 148 later converted to F-4N (see below). 29 loaned to USAF. 3 became YF-4J. 44 to QF-4B. 46 to RF-4B for Marines. First flight Mar 25, 1961. |
| **QF-4B** | (44) | Research remotely piloted vehicle (RPV) controlled from the air (DF-4B) or ground. Converted by Naval Air Development Centre, Warminster, Pennsylvania. |
| **RF-4B** (USMC) | 46 | Unarmed, equipped with cameras, radar, infra-red sensors, etc. Similar to RF-4C for USAF. First flight Mar 12, 1965. First delivery to USMC May 1965. |
| **F-4C** (F-110A) | 583 | USAF insisted that as much commonality with Navy aircraft as possible be retained – including folding wing and arrester gear – but specified dual controls, low-pressure tyres, boom rather than probe-and-drogue refuelling, different (cartridge) starting system, improved electronics including AN/APQ-100 radar, AN/ASN-48 inertial navigation system, anti-skid, ASJB-7 bombing system, wing bump for low-pressure tyres, etc. Go-ahead May 1962, first flight May 27, 1963. Four Sparrows, plus equipment for Sidewinders, Falcons, Shrikes, Walleyes, conventional bombs, etc. J79-GE-15 engines, 10,900/17,000lb thrust. 36 converted for Spanish use. Some Wild Weasel use. First to squadron, MacDill AFB, Tampa, Nov 11, 1963. |
| **RF-4C** | 505 | Unarmed. Films could be processed in flight and ejected at low altitude. Extended nose, tested on YRF-4Cs (62-12200, "the most reconfigured F-4", and 62-12201). APQ-99 forward-looking radar. APQ-102 side-looking radar. First flight May 18, 1964. |
| **F-4D** | 825 | First flight Dec 7, 1965. Improved air-to-ground capability. APQ-109 radar. ASN-63 inertial navigation. 32 to Imperial Iranian Air Force, 36 to Republic of Korea Air Force. First delivery to Warner Robin AFB, Georgia, Mar 10, 1966; last delivery Feb 28, 1969. |
| **F-4E** | 781 (USAF) 746 (foreign) | Vastly improved aircraft. J79-17 engines, 17,900lb thrust in afterburner. Built-in M161A 20mm Gatling. APQ-120 radar. ASG-26A lead-computing optical sight. Later fitted (after June 1972) with manoeuvring slats. YF-4E first flight Aug 7, 1965. F-4E prototype rollout Mar 1, 1967. Production F-4E first flight Jun 30, 1967. First delivery to TAC, Nellis AFB, Nevada, Oct 3, 1967. First flight manoeuvring-slat F-4E (71-0238) Feb 11, 1972, Job Dobronski pilot, Dick Richards RO. Used by Thunderbirds 1969-73. 204 to Israel, 10 to Germany, 177 to Iran, 56 to Greece, 87 to Turkey, 140 to Japan (138 built in Japan, 2 in US), 37 to Korea, 24 loaned temporarily to Australia and returned, 35 to Egypt. Aircraft built in Japan designated F-4EJ. EF-4E conversion of F-4E to Wild Weasel standard, 116 in programme. |
| **RF-4E** | 146 | Essentially the RF-4C nose mated to F-4E airframe for export. Wide variation in equipment depending upon country. 12 to Israel, 88 to Germany, 16 to Iran, 8 to Greece, 8 to Turkey, 14 to Japan. (Japanese versions designated RF-4EJ, built by MDC.) First flight Sept 15, 1970. |
| **F-4F** | 175 | Similar to F-4E but simplified maintenance and operations. Manoeuvring slats. First flight May 18, 1973. |

| Model | Quantity | Notes |
|---|---|---|
| F-4G (USN) | (12) | F-4B fitted with AN/ASW-21 two-way data-link communications system and approach power compensator for automatic carrier landing system. Later reconverted to F-4B. Flown by VF-213 from USS *Kitty Hawk*. Device later made standard in F-4J. First flight, BuNo 150481, Mar 20, 1963, Thomas E. Harris pilot, John J. Kiely RO. |
| F-4G (USAF) | (116) | F-4E converted to Wild Weasel role. Equipment includes ECM gear adapted continuously to meet new threats. Highly classified AN/APR-38 system substituted for internal gun. Advanced missile armament to suppress SAM/AAA. |
| F-4J | 522 | Advanced interceptor. J79-GE-10 engines, 17,900lb thrust. Ailerons droop, larger flaps, AWG-10 Doppler fire control, 30kVA alternators, fixed inboard leading-edge flap, 3,500-channel UHF, Martin-Baker zero-zero ejection seats. "Shoehorn" equipment. First flight F-4J May 27, 1966, Ray D. Hunt pilot, Charles E. Rosenmayer RO. First delivery to fleet, VF-101, Key West, Dec 21, 1966. Used by Blue Angels 1969-73. 260 to be reworked to F-4S (see below). Westinghouse APG-59 radar. |
| F-4K | 52 | 28 to RAF, 24 to Royal Navy. Modified F-4J with Rolls-Royce Spey 202/203 engines of 12,250/20,515lb thrust. 40-45% production in United Kingdom. Lengthened (extensible) nosewheel, inlet ducts widened, tailplane anhedral reduced, folding nose radome. Designated Phantom FG.1 in UK. First public flight June 28, 1966. First delivery April 25, 1968. |
| F-4M | 118 | RAF version of F-4K. Bigger brakes, low-pressure tyres, different equipment, new Ferranti/inertial nav, no extensible nose leg, stabiliser slot or drooped ailerons. Last delivery Aug 29, 1969. Both F-4K and F-4M sadly missed in Falklands War. RAF Phantoms now stationed at Port Stanley. |
| F-4N | (178) | Converted USN and USMC F-4Bs from Blocks 12 to 28. Service life extension programme. Prototype F-4N flown from NAS North Island Jun 4, 1972. First production F-4N (BuNo 150430) delivered to XV-4, NAS Pt Mugu, Feb 21, 1973, Lt-Cdr L. A. Lantzer pilot, Lt B. Lee RO. Naval Air Rework Facility completely overhauls, installs new equipment etc, for service life extension programme. Intended mainly for USNR squadrons. |
| F-4S | (260) | Similar programme by Naval Air Rework Centre for F-4Js. Scheduled for completion in 1984, gives increased strength, longer fatigue life, new mission equipment. All F-4S eventually to have slatted wing. |

# Appendix 2

This account of the highly successful Mission Bolo MiG-killing engagement of January 2, 1967, is taken from a working paper produced by the US Seventh Air Force Tactical Air Analysis Centre.

## I Introduction

1   The following description and analysis of Mission Bolo has been prepared to answer the many requests received for information on this very successful MiG engagement.

2   Mission Bolo was conducted by the 8th Tactical Fighter Wing base at Ubon. The mission was fragged for seven flights of 4 F-4Cs divided into 2 waves.

3   The success of Mission Bolo is largely attributable to:
   a   The overall planning and development of mission strategy and tactics, which accurately anticipated and fully exploited enemy reaction, and the attention to detail in the planning phase with particular focus on total force interaction in relation to both position and timing.
   b   An intensive training program for 8TFW combat aircrews which emphasized every facet of total mission to include missile capabilities, aircraft and missile procedures, MiG manueverability, radar search patterns, MiG identification, flight maneuvering and flight integrity, radio procedures, fuel management, tank jettison procedures etc.
   c   High degree of discipline, both ground and air, displayed by all participants.

## II Mission Narrative

1   The first wave consisted of Ford, Olds and Rambler Flights launched at 5-minute intervals starting at 1225 and the second wave was composed of Vespa, Plymouth, Lincoln and Tempest Flights launched at 1255. Mission proceeded smoothly as all flights arrived at the air refueling tracks at their scheduled times. Refueling support was termed excellent by all flights.

2   Tempest lead air-aborted due to refueling receptacle problems and Tempest 02 aborted sympathetically. Tempest 03 and 04 proceeded to point 2100/10500. Tempest 03 and 04 QRC-160 pods were found to be inoperative and decision not to penetrate was made.

3   Olds Flight proceeded as planned and arrived over target at 0700Z. On arriving over target no sighting was made. Flight entered orbit and made radar contact with MiGs, who were on a reciprocal heading and still in undercast. Olds Flight made a 180-degree turn and sighted MiGs breaking on top of undercast. Lead observed one MiG-21 moving into firing position. Flight broke into a defensive split. Two more MiGs were sighted in left turns. Lead fired 2 AIM-7s with interlocks in at the front MiG but lost radar lock due to minimum range as missiles were launched. One AIM-9 was fired as MiG ducked into undercast. Pressing left toward another MiG-21 topping undercast, lead performed a barrel roll and ended at MiG's 7 o'clock position; selected heat, pipper on MiG, excellent growl, two Sidewinders were launched. First missile tracked true and impacted MiG, blowing off left wing. MiG went into a violent spin and out of control. (Altitude 9000 feet, airspeed 525 kts, G load at launch 1.5, overtake slight, range to target 4500 feet.) Olds 02 sighted a MiG-21 at 10 o'clock, observed lead's launch of 2 AIM-7s. 02 locked on to the MiG-21 at about 30-40 degrees angle off and fired 2 AIM-7s. The second AIM-7 guided and impacted just forward of horizontal stabilizer of MiG-21. MiG exploded and fell in pieces (altitude 14,000 feet, attitude level, airspeed 560 kts, G load 1.5, slight overtake, range 6000-7000 feet). Olds 04 sighted MiG in the 6 o'clock position to Olds 01, 02 and 03 and turning outside their turn. 04 dropped down and outside, fired 1 AIM-9 missile which guided and produced ball of black smoke on impact. MiG-21 pitched violently down into undercast. (Altitude 14,000 feet, attitude level turn, airspeed 550 kts, G load 1.0, overtake, range 4500 feet, left turn of 10 degrees bank, loud Sidewinder tone.)

4   Rambler Flight ingressed as Ford Flight egressed. No MiGs sighted initially, but after one orbit 2 MiGs were observed at 2-3 o'clock low. Lead called in and launched 2 AIM-7s, one missile impacted, destroying the MiG. (Altitude 15,000 feet, attitude 10 degrees dive and 15 degrees left bank, rear-quarter attack, airspeed 580 kts, G load 1.5, range 6000 feet, 20-degree angle off, 150-foot overtake, radar lock, interlocks out.) A MiG-21 slid between lead and 02 with guns blazing. Lead and 02 broke up, left and then down. 02 came out of maneuver behind Rambler 04. 02 then turned inside 2 MiG-21s and fired 2 AIM-7s; the second AIM-7 impacted and destroyed the MiG-21. (Altitude 14,000-15,000 feet, attitude 30 degrees left bank, rear-quarter attack with barrel roll repositioning, airspeed 600 kts, G load 3-4, range 2½ miles, angle off 40 degrees, 200 kts overtake, full system.) Rambler Flight in right turn sighted MiGs, 2 o'clock low. Rambler 04 locked on with full system, interlocks out. Manually tracking the MiG, 04 fired two AIM-7s. Number 2 missile guided and detonated in the tailpipe of the MiG-21. There was an explosion followed by fireball. Another flight of MiGs was observed. Rambler 04, using full system with interlocks, tracked the MiGs manually and fired 2 AIM-7s. One missile was ineffective and the other was unobserved. After a high-G turn, 04 was again in position and fired 4 AIM-9s, with Number 1 AIM-9 detonating to the right and high of MiG tailpipe. Number 2 missile detonated right and low. Number 3 and 4 missiles were guiding when the flight broke right to avoid attacking MiG. A chute was observed for a probable kill. (Altitude 12,000 feet, airspeed Mach 1.1, attitude left bank, G unloaded, overtake 200 kts, range 1½ miles, angle off 25 degrees, Sidewinder tone loud.)

5   Lincoln and Vespa Flights entered the target area on time. No further sightings were made. All flights followed planned mission profiles with no further sightings.

### III Battle Summary

a   Olds 01, 02 and 04 were credited with one MiG-21 kill apiece. Olds Flight fired a total of 4 AIM-7E and 4 AIM-9B missiles. 4 MiG-21s were engaged by radar and this enabled Olds Flight to maintain an offensive advantage. The MiGs were engaged at a low altitude and airspeed, which also gave Olds Flight an advantage on total energy. Engagement altitude varied from 9,000 to 14,000 feet, which was a decided advantage for the F-4s as regards maneuverability.

b   Ford Flight was credited with 1 MiG-21 kill made by Ford 02. A total of 4 AIM-9Bs and 2 AIM-7Es were fired. 7 MiG-21s were engaged level, but offensive initiative was maintained through radar acquisition. Engagement altitudes varied from 10,000 to 12,000 feet, which was a decided advantage for the F-4C. Flight was in constant engagement with the enemy and many evasive maneuvers were necessary to avoid being hit. This situation differed from that of Olds and Rambler flights, which experienced a series of separate engagements.

c   Rambler Flight was credited with 3 MiG-21 confirmed kills made by Rambler 01, 02 and 04, with Rambler 03 and 04 credited with probable kills. A total of 12 AIM-7Es and 4 AIM-9Bs were fired. 7 MiG-21s were engaged at a lower energy spectrum (low airspeed and altitude) and the enemy was acquired visually. As with Olds Flight, Rambler Flight enjoyed a higher total energy capability (airspeed and altitude) advantage over the enemy. This correlates to the greater success in kills of Olds and Rambler flights over Ford Flight. Engagement altitudes varied from 12,000 to 15,000 feet.

# Appendix 3
# US Navy victories in Vietnam

| Date | Kill | Type of Aircraft | Crew Rank | Squadron and CV No | Pilot/NFO |
|---|---|---|---|---|---|
| 17 Jun 65 | MiG-17 | F-4 | CDR<br>LCDR | VF-21<br>41 | Louis C. Page<br>John Carl Smith Jr |
| 17 Jun 65 | MiG-17 | F-4 | LT<br>LCDR | VF-21<br>41 | Jack Ernest David Batson Jr<br>Robert Bartsch Doremus |
| 20 Jun 65 | MiG-17 | A-1 | LT | VA-25<br>41 | Clinton Bernard Johnson |
| 12 Jun 66 | MiG-17 | F-8 | CDR | VF-211<br>19 | Harold Lloyd Marr |
| 21 Jun 66 | MiG-17 | F-8 | LT | VF-211<br>19 | Eugene J. Chancy |
| 21 Jun 66 | MiG-17 | F-8 | LTJG | VF-211<br>19 | Phillip V. Vampatella |
| 13 Jul 66 | MiG-17 | F-4 | LT<br>LTJG | VF-161<br>64 | William M. McGuigan<br>Robert Matthew Fowler |
| 9 Oct 66 | MiG-21 | F-8 | CDR | VF-162<br>34 | Richard M. Bellinger |
| 9 Oct 66 | MiG-17 | A-1 | LTJG | VA-176<br>11 | William T. Patton |
| 24 Apr 67 | MiG-17 | F-4 | LT<br>LTJG | VF-114<br>63 | Hugh Dennis Wisely<br>Gareth L. Anderson |
| 24 Apr 67 | MiG-17 | F-4 | LCDR<br>ENS | VF-114<br>63 | Charles E. Southwick<br>James W. Laing |
| 1 May 67 | MiG-17 | F-8 | LCDR | VF-211<br>31 | Marshall Owen Wright |
| 1 May 67 | MiG-17 | A-4 | LCDR | VA-76<br>31 | Theodore Robert Swartz |
| 19 May 67 | MiG-17 | F-8 | CDR | VF-211<br>31 | Paul Howard Speer |
| 19 May 67 | MiG-17 | F-8 | LTJG | VF-211<br>31 | Joseph M. Shea |
| 19 May 67 | MiG-17 | F-8 | LCDR | VF-24<br>31 | Bobby Clyde Lee |
| 19 May 67 | MiG-17 | F-8 | LT | VF-24<br>31 | Phillip Ray Wood |
| 21 Jul 67 | MiG-17 | F-8 | CDR | VF-24<br>31 | Marion Howard Isaacks |
| 21 Jul 67 | MiG-17 | F-8 | LCDR | VF-24<br>31 | Robert Liston Kirkwood |
| 21 Jul 67 | MiG-17 | F-8 | LCDR | VF-211<br>31 | Ray George Hubbard Jr |

| Date | Kill | Type of Aircraft | Crew Rank | Squadron and CV No | Pilot/NFO |
|---|---|---|---|---|---|
| 10 Aug 67 | MiG-21 | F-4 | LT<br>ENS | VF-142<br>64 | Guy Herbert Freeborn<br>Robert Jurane Elliot |
| 10 Aug 67 | MiG-21 | F-4 | LCDR<br>LCDR | VF-142<br>64 | Robert C. Davis<br>Gayle Owen Elie |
| 26 Oct 67 | MiG-21 | F-4 | LTJG<br>LTJG | VF-143<br>64 | Robert Phillip Hickey Jr<br>Jeremy G. Morris |
| 30 Oct 67 | MiG-17 | F-4 | LCDR<br>LTJG | VF-142<br>64 | Eugene Patrick Lund<br>James Raymond Borst |
| 14 Dec 67 | MiG-17 | F-8 | LT | VF-162<br>34 | Richard E. Wyman |
| 26 Jun 68 | MiG-21 | F-8 | CDR | VF-51<br>31 | Lowell Richard Myers |
| 9 Jul 68 | MiG-17 | F-8 | LCDR | VF-191<br>14 | John Bennett Nichols III |
| 10 Jul 68 | MiG-21 | F-4 | LT<br>LT | VF-33<br>66 | Roy Cash Jr<br>Joseph Edward Kain Jr |
| 29 Jul 68 | MiG-17 | F-8 | CDR | VF-53<br>31 | Guy Cane |
| 1 Aug 68 | MiG-21 | F-8 | LT | VF-51<br>31 | Norman Kitchens McCoy Jr |
| 19 Sep 68 | MiG-21 | F-8 | LT | VF-111<br>11 | Anthony John Nargi |
| 28 Mar 70 | MiG-21 | F-4 | LT<br>LT | VF-142<br>64 | Jerome Eugene Beaulier<br>Steven John Barkley |
| 19 Jan 72 | MiG-21 | F-4 | LT<br>LTJG | VF-96<br>64 | Randall Harold Cunningham<br>William Patrick Driscoll |
| 6 Mar 72 | MiG-17 | F-4 | LT<br>LTJG | VF-111<br>43 | Garry Lee Weigand<br>William Cyrus Freckleton |
| 6 May 72 | MiG-17 | F-4 | LCDR<br>LT | VF-51<br>43 | Jerry Beaman Houston<br>Kevin Thomas Moore |
| 6 May 72 | MiG-21 | F-4 | LT<br>LTJG | VF-114<br>63 | Robert Garfield Hughes<br>Adolph Joseph Cruz |
| 6 May 72 | MiG-21 | F-4 | LCDR<br>LTJG | VF-114<br>63 | Kenneth William Pettigrew<br>Michael Joseph McCabe |
| 8 May 72 | MiG-17 | F-4 | LT<br>LTJG | VF-96<br>64 | Randall Harold Cunningham<br>William Patrick Driscoll |
| 10 May 72 | MiG-21 | F-4 | LT<br>LCDR | VF-92<br>64 | Curt Dose<br>James McDevitt |
| 10 May 72 | MiG-17 | F-4 | LT<br>LT | VF-96<br>64 | Michael J. Connelly<br>Thomas Joseph John Blonski |
| 10 May 72 | MiG-17 | F-4 | LT<br>LT | VF-51<br>43 | Roy Anthony Morris Jr<br>Kenneth Lee Cannon |
| 10 May 72 | MiG-17 | F-4 | LT<br>LT | VF-96<br>64 | Michael J. Connelly<br>Thomas Joseph John Blonski |
| 10 May 72 | MiG-17 | F-4 | LT<br>LTJG | VF-96<br>64 | Randall Harold Cunningham<br>William Patrick Driscoll |
| 10 May 72 | MiG-17 | F-4 | LT<br>LTJG | VF-96<br>64 | Randall Harold Cunningham<br>William Patrick Driscoll |

| Date | Kill | Type of Aircraft | Crew Rank | Squadron and CV No | Pilot/NFO |
|---|---|---|---|---|---|
| 10 May 72 | MiG-17 | F-4 | LT<br>LTJG | VF-96<br>64 | Randall Harold Cunningham<br>William Patrick Driscoll |
| 10 May 72 | MiG-17 | F-4 | LT<br>LTJG | VF-96<br>64 | Steven Collier Shoemaker<br>Keith Virgil Crenshaw |
| 18 May 72 | MiG-19 | F-4 | LT<br>LT | VF-161<br>41 | Henry Bartholomay<br>Oran Roger Brown |
| 18 May 72 | MiG-19 | F-4 | LT<br>LT | VF-161<br>41 | Patrick Earl Arwood<br>James Michael Bell |
| 23 May 72 | MiG-17 | F-4 | LCDR<br>LT | VF-161<br>41 | Ronald Eugene McKeown<br>John Clyde Ensch |
| 23 May 72 | MiG-17 | F-4 | LCDR<br>LT | VF-161<br>41 | Ronald Eugene McKeown<br>John Clyde Ensch |
| 11 Jun 72 | MiG-17 | F-4 | CDR<br>LT | VF-51<br>43 | Foster Schuler Teague<br>Ralph Marion Howell |
| 11 Jun 72 | MiG-17 | F-4 | LT<br>LT | VF-51<br>43 | William Winston Copeland Jr<br>Donald Robert Bouchoux |
| 21 June 72 | MiG-21 | F-4 | CDR<br>LT | VF-31<br>60 | Samuel Carson Flynn Jr<br>William Harrison John |
| 10 Aug 72 | MiG-21 | F-4 | LCDR<br>LTJG | VF-103<br>60 | Robert Eugene Tucker Jr<br>Stanley Bruce Edens |
| 11 Sep 72 | MiG-21 | F-4 | MAJ<br>CAPT | VMFA-333<br>66 | Lee T. Lassiter<br>John D. Cummings |
| 28 Dec 72 | MiG-21 | F-4 | LTJG<br>LTJG | VF-142<br>65 | Scott Harvey Davis<br>Geoffrey Hugh Ulrich |
| 12 Jan 73 | MiG-17 | F-4 | LT<br>LT | VF-161<br>41 | Victor Theodore Kovaleski<br>James Allen Wise |

Carrier names: 19 *Hancock*, 31 *Bon Homme Richard*, 34 *Oriskany*, 41 *Midway*, 43 *Coral Sea*, 60 *Saratoga*, 62 *Independence*, 63 *Kitty Hawk*, 64 *Constellation*, 65 *Enterprise*, 66 *America*, 11 *Intrepid*, 14 *Ticonderoga*.

# Appendix 4
# USAF victories in Vietnam

| Individual/Rank Crew position/Home town | USAF Sqn | Parent unit | Date | Credit | Type enemy aircraft | Type USAF aircraft | Radio callsign | Primary USAF weapon used |
|---|---|---|---|---|---|---|---|---|
| Anderson, Robert D, Maj AC, Tulsa, Oklahoma | 389 TFS | 366 TFW | 23 Apr 67 | 1.0 | MiG-21 | F-4C | Chicago 03 | AIM-7 |
| Anderson, Ronald C, Cpt P, Fairbanks, Alaska | 45 TFS | 2 AD | 10 Jul 65 | 1.0 | MiG-17 | F-4C | Unknown 04 | AIM-9 |
| Autrey, Daniel L, 1Lt WSO, Hialeah, Florida | 35 TFS | 388 TFW | 12 Sep 72 | 1.0 | MiG-21 | F-4E | Finch 03 | 20mm |
| Baily, Carl G, LtC AC, Denver, Colorado | 13 TFS | 432 TRW | 18 Jul 72<br>29 Jul 72 | 1.0<br>1.0 | MiG-21<br>MiG-21 | F-4D<br>F-4D | Snug 01<br>Cadillac 01 | AIM-9<br>AIM-7 |
| Baker, Doyle D, Cpt (USMC) AC | 13 TFS | 432 TRW | 17 Dec 67 | 1.0 | MiG-17 | F-4D | Gambit 03 | AIM-4 |
| Bakke, Samuel O, Maj AC, Fort Morgan, Colorado | 480 TFS | 366 TFW | 14 May 67 | 1.0 | MiG-17 | F-4C | Elgin 01 | AIM-7 |
| Barton, Charles D, Cpt AC, Greenville, SC | 34 TFS | 388 TFW | 06 Oct 72 | 0.5 | MiG-19 | F-4E | Eagle 04 | Manoeuvring |
| Basel, Gene I, Cpt P, Lakeside, California | 354 TFS | 355 TFW | 27 Oct 67 | 1.0 | MiG-17 | F-105D | Bison 02 | 20mm |
| Battista, Robert B, 1Lt P, Montgomery, Alabama | 433 TFS | 8 TFW | 06 Feb 68 | 1.0 | MiG-21 | F-4D | Buick 04 | AIM-7 |
| Beatty, James M Jr, Cpt AC, Eau Claire, Pa | 35 TFS | 366 TFW | 23 May 72 | 1.0 | MiG-21 | F-4E | Balter 03 | 20mm |
| Beckers, Lyle L, LtC AC, Gregory, SD | 35 TFS<br>35 TFS | 366 TFW<br>388 TFW | 23 May 72<br>12 Sep 72 | 1.0<br>1.0 | MiG-19<br>MiG-21 | F-4E<br>F-4E | Balter 01<br>Finch 01 | AIM-7<br>AIM-9/20mm |
| Bell, James R, 1Lt WSO, Springfield, Ohio | 555 TFS | 432 TRW | 11 May 72 | 1.0 | MiG-21 | F-4D | Gopher 02 | AIM-7 |
| Bettine, Frank J, 1Lt WSO, Hartshorne, Oklahoma | 4 TFS | 366 TFW | 19 Aug 72 | 1.0 | MiG-21 | F-4E | Pistol 03 | AIM-7 |
| Bever, Michael R, 1Lt P, Kansas City, Missouri | 433 TFS | 8 TFW | 13 May 67 | 1.0 | MiG-17 | F-4C | Harpoon 03 | AIM-7 |
| Binkley, Eldon D, 1Lt WSO, Winston-Salem, NC | 555 TFS | 432 TRW | 22 Dec 72 | 1.0 | MiG-21 | F-4D | Bucket 01 | Manoeuvring |
| Blake, Robert E, Cpt AC, Presque Isle, Maine | 555 TFS | 8 TFW | 23 Apr 66 | 1.0 | MiG-27 | F-4C | Unknown 04 | AIM-7 |
| Blank, Kenneth T, Maj P, Franklin, Nebraska | 34 TFS | 388 TFW | 18 Aug 66 | 1.0 | MiG-17 | F-105D | Honda 02 | 20mm |
| Bleakley, Robert A, 1Lt P, Cedar Rapids, Iowa | 555 TFS | 8 TFW | 29 Apr 66 | 1.0 | MiG-17 | F-4C | Unknown 01 | Manoeuvring |
| Bogoslofski, Bernard J, Maj AC, Granby, Connecticut | 433 TFS | 8 TFW | 03 Jan 68 | 1.0 | MiG-17 | F-4D | Tampa 01 | 20mm |
| Boles, Robert H, Cpt AC, Lexington, SC | 433 TFS | 8 TFW | 06 Feb 68 | 1.0 | MiG-21 | F-4D | Buick 04 | AIM-7 |
| Bongartz, Theodore R, 1Lt P, Catonsville, Maryland | 433 TFS | 8 TFW | 24 Oct 67 | 1.0 | MiG-21 | F-4D | Buick 01 | 20mm |

| Individual/Rank Crew position/Home town | USAF Sqn | Parent unit | Date | Credit | Type enemy aircraft | Type USAF aircraft | Radio callsign | Primary USAF weapon used |
|---|---|---|---|---|---|---|---|---|
| Brestel, Max C, Cpt P, Chappell, Nebraska | 354 TFS | 355 TFW | 10 Mar 67 10 Mar 67 | 1.0 1.0 | MiG-17 MiG-17 | F-105D F-105D | Kangaroo 03 Kangaroo 03 | 20mm 20mm |
| Brown, Frederick W, Maj WSO, Grand View, Wash | 523 TFS | 432 TRW | 15 Oct 72 | 1.0 | MiG-21 | F-4D | Chevy 01 | AIM-9 |
| Brunson, Cecil H, 1Lt WSO, Memphis, Tennessee | 34 TFS | 388 TFW | 06 Oct 72 | 0.5 | MiG-19 | F-4E | Eagle 03 | Manoeuvring |
| Brunson, James E, LtC AC, Eddyville, Kentucky | 555 TFS | 432 TRW | 22 Dec 72 | 1.0 | MiG-21 | F-4D | Buick 01 | AIM-7 |
| Buttell, Duane A Jr, 1Lt P, Chillicothe, Illinois | 480 TFS | 35 TFW | 14 Jul 66 | 1.0 | MiG-21 | F-4C | Unknown 01 | AIM-9 |
| Cameron, Max F, Cpt AC, Stanford, NC | 555 TFS | 8 TFW | 23 Apr 66 | 1.0 | MiG-17 | F-4C | Unknown 04 | AIM-9 |
| Cary, Lawrence E, 1Lt P, Pawnee City, Nebraska | 433 TFS | 8 TFW | 02 Jan 67 | 1.0 | MiG-21 | F-4C | Rambler 02 | AIM-7 |
| Cherry, Edward D, Maj AC, Marietta, Georgia | 13 TFS | 432 TRW | 16 Apr 72 | 1.0 | MiG-21 | F-4D | Basco 03 | AIM-7 |
| Christiansen, Von R, LtC AC, Seattle, Washington | 469 TFS | 388 TFW | 21 Jun 72 | 1.0 | MiG-21 | F-4E | Iceman 03 | AIM-9 |
| Clark, Arthur C, Cpt P, McAllen, Texas | 45 TFS | 2 AD | 10 Jul 65 | 1.0 | MiG-17 | F-4C | Unknown 03 | AIM-9 |
| Clifton, Charles, 1Lt P, Fort Wayne, Indiana | 555 TFS | 8 TFW | 02 Jan 67 | 1.0 | MiG-21 | F-4C | Olds 01 | AIM-9 |
| Clouser, Gordon L, Maj AC, Norman, Oklahoma | 34 TFS | 388 TFW | 06 Oct 72 | 0.5 | MiG-19 | F-4E | Eagle 03 | Manoeuvring |
| Cobb, Larry D, Cpt AC, Lambert, Missouri | 555 TFS | 8 TFW | 26 Oct 67 | 1.0 | MiG-17 | F-4D | Ford 04 | AIM-4 |
| Coe, Richard E, Cpt AC, East Orange, NJ | 34 TFS | 388 TFW | 05 Oct 72 | 1.0 | MiG-21 | F-4E | Robin 01 | AIM-7 |
| Combies, Philip P, Maj AC, Norwich, Connecticut | 433 TFS | 8 TFW | 02 Jan 67 20 May 67 | 1.0 1.0 | MiG-21 MiG-17 | F-4C F-4C | Rambler 04 Ballot 01 | AIM-7 AIM-9 |
| Cooney, James, P, LtC WSO, Newburgh, New York | 555 TFS | 432 TRW | 12 May 72 | 1.0 | MiG-19 | F-4D | Harlow 02 | AIM-7 |
| Couch, Charles W, Cpt P, Caseyville, Illinois | 354 TFS | 355 TFW | 13 May 67 | 1.0 | MiG-17 | F-105D | Chevrolet 03 | 20mm |
| Craig, James T Jr, Cpt AC, Abilene, Texas | 480 TFS | 366 TFW | 14 May 67 | 1.0 | MiG-17 | F-4C | Speedo 03 | 20mm |
| Crews, Barton P, Maj AC, Fort Lauderdale, Fla | 13 TFS | 432 TRW | 08 May 72 | 1.0 | MiG-19 | F-4D | Galore 03 | AIM-7 |
| Croker, Stephen B, 1Lt P, Middletown, Delaware | 433 TFS | 8 TFW | 20 May 67 20 May 67 | 1.0 1.0 | MiG-17 MiG-17 | F-4C F-4C | Tampa 01 Tampa 01 | AIM-7 AIM-9 |
| Dalton, William M, Maj P, Stephens City, Virginia | 333 TFS | 355 TFW | 19 Dec 67 | 0.5 | MiG-17 | F-105F | Otter 02 | 20mm |
| De Bellevue, Charles B, Cpt WSO, Lafayette, Louisiana | 555 TFS | 432 TRW | 10 May 72 08 Jul 72 08 Jul 72 28 Aug 72 09 Sep 72 09 Sep 72 | 1.0 1.0 1.0 1.0 1.0 1.0 | MiG-21 MiG-21 MiG-21 MiG-21 MiG-19 MiG-19 | F-4D F-4E F-4E F-4D F-4D F-4D | Oyster 03 Paula 01 Paula 01 Buick 01 Olds 01 Olds 01 | AIM-7 AIM-7 AIM-7 AIM-7 AIM-9 AIM-9 |
| DeMuth, Stephen H, 1Lt P, Medina, Ohio | 480 TFS | 366 TFW | 14 May 67 | 1.0 | MiG-17 | F-4C | Speedo 01 | 20mm |
| Dennis, Arthur F, LtC P, Sherman, Texas | 357 TFS | 355 TFW | 28 Apr 67 | 1.0 | MiG-17 | F-105D | Atlanta 01 | 20mm |
| Dickey, Roy S, Maj P, Ashland, Kansas | 469 TFS | 388 TFW | 04 Dec 66 | 1.0 | MiG-17 | F-105D | Elgin 04 | 20mm |

| Individual/Rank Crew position/Home town | USAF Sqn | Parent unit | Date | Credit | Type enemy aircraft | Type USAF aircraft | Radio callsign | Primary USAF weapon used |
|---|---|---|---|---|---|---|---|---|
| Diehl, William C, 1Lt WSO, Tampa, Florida | 34 TFS | 388 TFW | 15 Oct 72 | 1.0 | MiG-21 | F-4E | Parrot 03 | AIM-9 |
| Dilger, Robert G, Maj AC, Tampa, Florida | 390 TFS | 366 TFW | 01 May 67 | 1.0 | MiG-17 | F-4C | Stinger 01 | Manoeuvring |
| Dowell, William B D, Cpt AC, Tampa, Florida | 555 TFS | 8 TFW | 29 Apr 66 | 1.0 | MiG-17 | F-4C | Unknown 03 | AIM-9 |
| Drew, Philip M, Cpt P, Alexandria, Louisiana | 357 TFS | 355 TFW | 19 Dec 67 | 1.0 | MiG-17 | F-105F | Otter 03 | 20mm |
| Dubler, John E, Cpt WSO, Omaha, Nebraska | 555 TFS | 432 TRW | 28 Dec 72 | 1.0 | MiG-21 | F-4D | List 01 | AIM-7 |
| Dudley, Wilbur R, Maj AC, Alamogordo, NM | 390 TFS | 35 TFWE | 12 May 66 | 1.0 | MiG-17 | F-4C | Unknown 03 | AIM-9 |
| Dunnegan, Clifton P Jr, 1Lt P, Winston-Salem, NC | 433 TFS | 8 TFW | 02 Jan 67 | 1.0 | MiG-21 | F-4C | Rambler 01 | AIM-7 |
| Dutton, Lee R, 1Lt P, Wyoming, Illinois | 433 TFS | 8 TFW | 02 Jan 67 | 1.0 | MiG-21 | F-4C | Rambler 04 | AIM-7 |
| Eaves, Stephen D, Cpt WSO, Honolulu, Hawaii | 555 TFS | 432 TRW | 10 May 72 | 1.0 | MiG-21 | F-4D | Oyster 02 | AIM-7 |
| Eskew, William E, Cpt P, Boonville, Indiana | 354 TFS | 355 TFW | 19 Apr 67 | 1.0 | MiG-17 | F-105D | Panda 01 | 20mm |
| Ettel, Michael J, LtCdr (USN) WSO, St Paul, Minn | 58 TFS | 432 TRW | 12 Aug 72 | 1.0 | MiG-21 | F-4E | Dodge 01 | AIM-7 |
| Evans, Robert E, 1Lt P, Haina, Hawaii | 555 TFS | 8 TFW | 23 Apr 66 | 1.0 | MiG-17 | F-4C | Unknown 03 | AIM-9 |
| Feighny, James P Jr, 1Lt P, Laramie, Wyoming | 435 TFS | 8 TFW | 14 Feb 68 | 1.0 | MiG-17 | F-4D | Killer 01 | AIM-7 |
| Feinstein, Jeffrey S, Cpt WSO, East Troy, Wisconsin | 13 TFS | 432 TRW | 16 Apr 72 | 1.0 | MiG-21 | F-4D | Basco 03 | AIM-7 |
|  |  |  | 31 May 72 | 1.0 | MiG-21 | F-4E | Gopher 03 | AIM-9 |
|  |  |  | 18 Jul 72 | 1.0 | MiG-21 | F-4D | Snug 01 | AIM-9 |
|  |  |  | 29 Jul 72 | 1.0 | MiG-21 | F-4D | Cadillac 01 | AIM-7 |
|  |  |  | 13 Oct 72 | 1.0 | MiG-21 | F-4D | Olds 01 | AIM-7 |
| Frye, Wayne T, LtC AC, Maysville, Kentucky | 555 TFS | 432 TRW | 12 May 72 | 1.0 | MiG-19 | F-4D | Harlow 02 | AIM-7 |
| Gast, Philip C, LtC P, Ewing, Missouri | 354 TFS | 355 TFW | 13 May 67 | 1.0 | MiG-17 | F-105D | Chevrolet 01 | 20mm |
| George, S W, 1Lt P, Canadian, Oklahoma | 555 TFS | 8 TFW | 23 Apr 66 | 1.0 | MiG-17 | F-4C | Unknown 04 | AIM-7 |
| Gilmore, Paul J, Maj AC, Alamogordo, NM | 480 TFS | 35 TFW | 26 Apr 66 | 1.0 | MiG-21 | F-4C | Unknown 01 | AIM-9 |
| Glynn, Lawrence J Jr, 1Lt AC, Arlington, Massachusetts | 433 TFS | 8 TFW | 02 Jan 67 | 1.0 | MiG-21 | F-4C | Rambler 02 | AIM-7 |
| Golberg, Lawrence H, Cpt AC, Duluth, Minnesota | 555 TFS | 8 TFW | 30 Apr 66 | 1.0 | MiG-17 | F-4C | Unknown 04 | AIM-9 |
| Gordon, William S III, Cpt AC, Wethersfield, Conn | 555 TFS | 8 TFW | 26 Oct 67 | 1.0 | MiG-17 | F-4D | Ford 03 | AIM-7 |
| Gossard, Halbert E, 1Lt P, Oklahoma City, Oklahoma | 555 TFS | 8 TFW | 29 Apr 66 | 1.0 | MiG-17 | F-4C | Unknown 03 | AIM-9 |
| Graham, James L, Maj EWO, Lancaster, Pennsylvania | 333 TFS | 355 TFW | 19 Dec 67 | 0.5 | MiG-17 | F-105F | Otter 02 | 20mm |
| Griffin, Thomas M, 1Lt WSO, New Orleans, Louisiana | 35 TFS | 388 TFW | 12 Sep 72 | 1.0 | MiG-21 | F-4E | Finch 01 | AIM-9/20mm |
| Gullick, Francis M, Cpt P, Albuquerque, New Mexico | 555 TFS | 8 TFW | 05 Jun 67 | 1.0 | MiG-17 | F-4D | Drill 01 | AIM-7 |

| Individual/Rank Crew position/Home town | USAF Sqn | Parent unit | Date | Credit | Type enemy aircraft | Type USAF aircraft | Radio callsign | Primary USAF weapon used |
|---|---|---|---|---|---|---|---|---|
| Haeffner, Fred A, LtC AC, Fargo, North Dakota | 433 TFS | 8 TFW | 13 May 67 | 1.0 | MiG-17 | F-4C | Harpoon 03 | AIM-7 |
| Handley, Philip W, Maj AC, Wellington, Texas | 58 TFS | 432 TRW | 02 Jun 72 | 1.0 | MiG-19 | F-4E | Brenda 01 | 20mm |
| Harden, Kaye M, Maj WSO, Jacksonville, Florida | 469 TFS | 388 TFW | 21 Jun 72 | 1.0 | MiG-21 | F-4E | Iceman 03 | AIM-9 |
| Hardgrave, Gerald D, 1Lt P, Jackson, Tennessee | 555 TFS | 8 TFW | 30 Apr 66 | 1.0 | MiG-17 | F-4C | Unknown 04 | AIM-9 |
| Hardy, Richard F, Cpt AC, Chicago, Illinois | 4 TFS | 366 TFW | 08 Jul 72 | 1.0 | MiG-21 | F-4E | Brenda 03 | AIM-7 |
| Hargrove, James A Jr, Maj AC, Garden City Beach, SC | 480 TFS | 366 TFW | 14 May 67 | 1.0 | MiG-17 | F-4C | Speedo 01 | 20mm |
| Hargrove, William S, 1Lt WSO, Harlingen, Texas | 555 TFS | 432 TRW | 09 Sep 72 16 Sep 72 | 1.0 1.0 | MiG-21 MiG-21 | F-4D F-4E | Olds 03 Chevy 03 | 20mm AIM-9 |
| Hendrickson, James L, Cpt WSO, Columbus, Ohio | 307 TFS | 432 TRW | 15 Oct 72 | 1.0 | MiG-21 | F-4E | Buick 03 | 20mm |
| Higgins, Harry E, Maj P, Alexandria, Indiana | 357 TFS | 355 TFW | 28 Apr 67 | 1.0 | MiG-17 | F-105D | Spitfire 01 | 20mm |
| Hill, Robert G, Cpt AC, Tucson, Arizona | 13 TFS | 432 TRW | 05 Feb 68 | 1.0 | MiG-21 | F-4D | Gambit 03 | AIM-4 |
| Hirsch, Thomas M, Maj AC, Rockford, Illinois | 555 TFS | 8 TFW | 06 Jan 67 | 1.0 | MiG-21 | F-4C | Crab 02 | AIM-7 |
| Hodgson, Leigh A, 1Lt WSO, Kingsport, Pennsylvania | 555 TFS | 432 TRW | 01 Mar 72 | 1.0 | MiG-21 | F-4D | Falcon 54 | AIM-7 |
| Holcombe, Kenneth E, Cpt AC, Detroit, Michigan | 45 TFS | 2 AD | 10 Jul 65 | 1.0 | MiG-17 | F-4C | Unknown 03 | AIM-9 |
| Holtz, Robert L, Maj AC, Milwaukee, Wisconsin | 34 TFS | 388 TFW | 15 Oct 72 | 1.0 | MiG-21 | F-4E | Parrot 03 | AIM-9 |
| Howerton, Rex D, Maj AC, Oklahoma City, Oklahoma | 555 TFS | 8 TFW | 14 Feb 68 | 1.0 | MiG-17 | F-4D | Nash 03 | 20mm |
| Howman, Paul D, Cpt AC, Wooster, Ohio | 4 TFS | 432 TRW | 08 Jan 73 | 1.0 | MiG-21 | F-4D | Crafty 01 | AIM-7 |
| Huneke, Bruce V, 1Lt P, Hanford, California | 13 TFS | 432 TRW | 05 Feb 68 | 1.0 | MiG-21 | F-4D | Gambit 03 | AIM-4 |
| Hunt, Jack W, Maj P, Freeport, Texas | 354 TFS | 355 TFW | 19 Apr 67 | 1.0 | MiG-17 | F-105D | Nitro 01 | 20mm |
| Huskey, Richard L, Cpt P, Cleveland, Tennessee | 433 TFS | 8 TFW | 03 Jan 68 | 1.0 | MiG-17 | F-4D | Tampa 01 | 20mm |
| Huwe, John F, Cpt WSO, Dell Rapids, SD | 35 TFS | 366 TFW | 23 May 72 | 1.0 | MiG-19 | F-4E | Balter 01 | AIM-7 |
| Imaye, Stanley M, Cpt WSO, Honolulu, Hawaii | 4 TFS | 366 TFW | 29 Jul 72 | 1.0 | MiG-21 | F-4E | Pistol 01 | AIM-7 |
| Jameson, Jerry W, 1Lt AC, Middletown, Indiana | 555 TFS | 8 TFW | 16 Sep 66 | 1.0 | MiG-17 | F-4C | Unknown 04 | AIM-9 |
| Janca, Robert D, Maj AC, Hampton, Virginia | 389 TFS | 366 TFW | 20 May 67 | 1.0 | MiG-21 | F-4C | Elgin 01 | AIM-9 |
| Jasperson, Robert H, Cpt WSO, Minneapolis, Minn | 35 TFS | 388 TFW | 08 Oct 72 | 1.0 | MiG-21 | F-4E | Lark 01 | 20mm |
| Johnson, Harold E, Cpt EWO, Blakeberg, Iowa | 357 TFS | 355 TFW | 19 Apr 67 | 1.0 | MiG-17 | F-105F | Kingfish 01 | 20mm |
| Jones, Keith W Jr, Cpt WSO, Glen Ellyn, Illinois | 13 TFS | 432 TRW | 08 May 72 | 1.0 | MiG-19 | F-4D | Galore 03 | AIM-7 |

| Individual/Rank Crew position/Home town | USAF Sqn | Parent unit | Date | Credit | Type enemy aircraft | Type USAF aircraft | Radio callsign | Primary USAF weapon used |
|---|---|---|---|---|---|---|---|---|
| Keith, Larry R, Cpt<br>AC, Peoria, Illinois | 555 TFS | 8 TFW | 29 Apr 66 | 1.0 | MiG-17 | F-4C | Unknown 01 | Manoeuvring |
| Kirk, William L, Maj<br>AC, Rayville, Louisiana | 433 TFS | 8 TFW | 13 May 67<br>24 Oct 67 | 1.0<br>1.0 | MiG-17<br>MiG-21 | F-4C<br>F-4D | Harpoon 01<br>Buick 01 | AIM-9<br>20mm |
| Kittinger, Joseph W Jr, LtC<br>AC, Orlando, Florida | 555 TFS | 432 TRW | 01 Mar 72 | 1.0 | MiG-21 | F-4D | Falcon 54 | AIM-7 |
| Kjer, Fred D, Cpt<br>P, Allen, Nebraska | 389 TFS | 366 TFW | 23 Apr 67 | 1.0 | MiG-21 | F-4C | Chicago 03 | AIM-7 |
| Klause, Klaus J, 1Lt<br>P, Franklin, Pennsylvania | 480 TFS | 366 TFW | 05 Nov 66 | 1.0 | MiG-21 | F-4C | Opal 02 | AIM-9 |
| Krieps, Richard N, 1Lt<br>P, Chesterton, Indiana | 480 TFS | 35 TFW | 14 Jul 66 | 1.0 | MiG-21 | F-4C | Unknown 02 | AIM-9 |
| Kringelis, Imants, 1Lt<br>P, Lake Zurich, Illinois | 390 TFS | 35 TFW | 12 May 66 | 1.0 | MiG-17 | F-4C | Unknown 03 | AIM-9 |
| Kullman, Lawrence W, 1Lt<br>WSO, Hartley, Delaware | 4 TFS | 432 TRW | 08 Jan 73 | 1.0 | MiG-21 | F-4D | Crafty 01 | AIM-7 |
| Kuster, Ralph L Jr, Maj<br>P, St Louis, Missouri | 13 TFS | 388 TFW | 03 Jun 67 | 1.0 | MiG-17 | F-105D | Hambone 02 | 20mm |
| Lafever, William D, 1Lt<br>P, Losantville, Indiana | 555 TFS | 8 TFW | 04 May 67 | 1.0 | MiG-21 | F-4C | Flamingo 01 | AIM-9 |
| Lafferty, Daniel L, 1Lt<br>P, Eddyville, Illinois | 433 TFS | 8 TFW | 20 May 67 | 1.0 | MiG-17 | F-4C | Ballot 01 | AIM-9 |
| Lambert, Robert W, Cpt<br>P, Virginia Beach, Virginia | 480 TFS | 366 TFW | 14 May 67 | 1.0 | MiG-17 | F-4C | Elgin 01 | AIM-7 |
| Lang, Alfred E Jr, LtC<br>AC, East Orange, NJ | 435 TFS | 8 TFW | 12 Feb 68 | 1.0 | MiG-21 | F-4D | Buick 01 | AIM-7 |
| Latham, Wilbur J Jr, 1Lt<br>AC, Eagle Grove, Iowa | 480 TFS | 366 TFW | 05 Nov 66 | 1.0 | MiG-21 | F-4C | Opal 02 | AIM-9 |
| Lavoy, Alan A, Cpt<br>P, Norwalk, Connecticut | 555 TFS | 8 TFW | 26 Oct 67 | 1.0 | MiG-17 | F-4D | Ford 04 | AIM-4 |
| Leonard, Bruce G Jr, Cpt<br>AC, Greensboro, NC | 13 TFS | 432 TRW | 31 May 72 | 1.0 | MiG-21 | F-4E | Gopher 03 | AIM-9 |
| Lesan, Thomas C, Cpt<br>P, Lebanon, Ohio | 333 TFS | 355 TFW | 30 Apr 67 | 1.0 | MiG-17 | F-105D | Rattler 01 | 20mm |
| Lewinski, Paul T, Cpt<br>WSO, Schenectady, New York | 4 TFS | 366 TFW | 08 Jul 72 | 1.0 | MiG-21 | F-4E | Brenda 03 | AIM-7 |
| Locher, Roger C, 1Lt/Cpt<br>WSO, Sabetha, Kansas | 555 TFS | 432 TRW | 21 Feb 72<br>08 May 72<br>10 May 72 | 1.0<br>1.0<br>1.0 | MiG-21<br>MiG-21<br>MiG-21 | F-4D<br>F-4D<br>F-4D | Falcon 62<br>Oyster 01<br>Oyster 01 | AIM-7<br>AIM-7<br>AIM-7 |
| Lodge, Robert A, Maj<br>AC, Columbus, Ohio | 555 TFS | 432 TRW | 21 Feb 72<br>08 May 72<br>10 May 72 | 1.0<br>1.0<br>1.0 | MiG-21<br>MiG-21<br>MiG-21 | F-4D<br>F-4D<br>F-4D | Falcon 62<br>Oyster 01<br>Oyster 01 | AIM-7<br>AIM-7<br>AIM-7 |
| Logeman, John D Jr, Cpt<br>AC, Fond Du Lac, Wisconsin | 555 TFS | 8 TFW | 26 Oct 67 | 1.0 | MiG-17 | F-4D | Ford 01 | AIM-7 |
| Lucas, Jon I, Maj<br>AC, Steubenville, Ohio | 34 TFS | 388 TFW | 02 Sep 72 | 1.0 | MiG-19 | F-4E | Eagle 03 | AIM-7 |
| Maas, Stuart W, Cpt<br>WSO, Williamsburg, Ohio | 13 TFS | 432 TRW | 16 Apr 72 | 1.0 | MiG-21 | F-4D | Basco 01 | AIM-7 |
| Madden, John A Jr, Cpt<br>AC, Jackson, Mississippi | 555 TFS | 432 TRW | 09 Sept 72<br>09 Sep 72<br>12 Oct 72 | 1.0<br>1.0<br>1.0 | MiG-19<br>MiG-19<br>MiG-21 | F-4D<br>F-4D<br>F-4D | Olds 01<br>Olds 01<br>Vega 01 | AIM-9<br>AIM-9<br>Manoeuvring |
| Mahaffey, Michael J, Cpt<br>AC, Patterson, California | 469 TFS | 388 TFW | 12 Sep 72 | 1.0 | MiG-21 | F-4D | Robin 02 | AIM-9 |

| Individual/Rank Crew position/Home town | USAF Sqn | Parent unit | Date | Credit | Type enemy aircraft | Type USAF aircraft | Radio callsign | Primary USAF weapon used |
|---|---|---|---|---|---|---|---|---|
| Malloy, Douglas G, 1Lt WSO, Dayton, Ohio | 35 TFS | 388 TFW | 02 Sep 72 | 1.0 | MiG-19 | F-4E | Eagle 03 | AIM-7 |
| Markle, John D, 1Lt AC, Hutchinson, Kansas | 555 TFS | 432 TRW | 10 May 72 | 1.0 | MiG-21 | F-4D | Oyster 02 | AIM-7 |
| Martin, Ronald G, 1Lt AC, Lake Villa, Illinois | 480 TFS | 35 TFW | 14 Jul 66 | 1.0 | MiG-21 | F-4C | Unknown 02 | AIM-9 |
| Massen, Mark A, Cpt WSO, Downey, California | 336 TFS | 8 TFW | 15 Aug 72 | 1.0 | MiG-21 | F-4E | Date 04 | AIM-7 |
| McCoy, Frederick E II, 1Lt P, Sheboygen, Wisconsin | 555 TFS | 8 TFW | 26 Oct 67 | 1.0 | MiG-17 | F-4D | Ford 01 | AIM-7 |
| McCoy, Ivy J Jr, Maj AC, Baton Rouge, Louisiana | 523 TFS | 432 TRW | 15 Oct 72 | 1.0 | MiG-21 | F-4D | Chevy 01 | AIM-9 |
| McKee, Harry L Jr, Maj AC, Austin, Texas | 555 TFS | 432 TRW | 28 Dec 72 | 1.0 | MiG-21 | F-4D | List 01 | AIM-7 |
| McKinney, George H Jr, 1Lt P, Bessemer, Alabama | 435 TFS | 8 TFW | 06 Nov 67 06 Nov 67 19 Dec 67 | 1.0 1.0 0.5 | MiG-17 MiG-17 MiG-17 | F-4D F-4D F-4D | Sapphire 01 Sapphire 01 Nash 01 | 20mm 20mm 20mm |
| Monsees, James H, 1Lt P, Santa Clara, California | 555 TFS | 8 TFW | 26 Oct 67 | 1.0 | MiG-17 | F-4D | Ford 03 | AIM-7 |
| Moore, Albert E, A1C G, San Bernardino, California | | 307 SW | 24 Dec 72 | 1.0 | MiG-21 | B-52D | Ruby III | .50 calibre |
| Moore, Joseph D, Maj AC, Spartanburg, SC | 435 TFS | 8 TFW | 19 Dec 67 | 0.5 | MiG-17 | F-4D | Nash 01 | 20mm |
| Moore, Rolland W Jr, Maj AC, Barberton, Ohio | 389 TFS | 366 TFW | 25 Apr 67 | 1.0 | MiG-21 | F-4C | Cactus 01 | AIM-7 |
| Moss, Randy P, 1Lt P, Great Falls, SC | 435 TFS | 8 TFW | 12 Feb 68 | 1.0 | MiG-21 | F-4D | Buick 01 | AIM-7 |
| Muldoon, Michael D, 1Lt P, Perry, New York | 435 TFS | 8 TFW | 03 Jan 68 | 1.0 | MiG-17 | F-4D | Olds 01 | AIM-4 |
| Murray, James E III, 1Lt P, McKeesport, Pennsylvania | 555 TFS | 8 TFW | 02 Jan 67 | 1.0 | MiG-21 | F-4C | Olds 04 | AIM-9 |
| Nichols, Stephen E, Cpt AC, Durham, NC | 555 TRS | 432 TRW | 11 May 72 | 1.0 | MiG-21 | F-4D | Gopher 02 | AIM-7 |
| Null, James C, Cpt AC, Oklahoma City, Oklahoma | 523 TFS | 432 TRW | 16 Apr 72 | 1.0 | MiG-21 | F-4D | Papa 03 | AIM-7 |
| Olds, Robin, Col AC, Washington, DC | 555 TFS 433 TFS | 8 TFW 8 TFW | 02 Jan 67 04 May 67 20 May 67 20 May 67 | 1.0 1.0 1.0 1.0 | MiG-21 MiG-21 MiG-17 MiG-17 | F-4C F-4C F-4C F-4C | Olds 01 Flamingo 01 Tampa 01 Tampa 01 | AIM-9 AIM-9 AIM-7 AIM-9 |
| Olmsted, Frederick S Jr, Cpt AC, San Diego, California | 13 TFS | 432 TRW | 30 Mar 72 16 Apr 72 | 1.0 1.0 | MiG-21 MiG-21 | F-4D F-4D | Papa 01 Basco 01 | AIM-7 AIM-7 |
| Osborne, Carl D, Maj P, Potlatch, Idaho | 333 TFS | 355 TFW | 13 May 67 | 1.0 | MiG-17 | F-105D | Random 03 | AIM-9 |
| Pankhurst, John E, Cpt P, Midland, Michigan | 480 TFS | 366 TFW | 05 Jun 67 | 1.0 | MiG-17 | F-4C | Oakland 01 | 20mm |
| Pardo, John R, Maj AC, Hearne, Texas | 433 TFS | 8 TFW | 20 May 67 | 1.0 | MiG-17 | F-4C | Tampa 03 | AIM-9 |
| Pascoe, Richard M, Cpt AC, Lakeside, California | 555 TFS | 8 TFW | 06 Jan 67 05 Jun 67 | 1.0 1.0 | MiG-21 MiG-17 | F-4C F-4C | Crab 01 Chicago 02 | AIM-7 AIM-9 |
| Pettit, Lawrence H, Cpt WSO, Jackson Heights, NY | 555 TFS | 432 TRW | 31 May 72 12 Oct 72 | 1.0 1.0 | MiG-21 MiG-21 | F-4D F-4D | Icebag 01 Vega 01 | AIM-7 Manoeuvring |
| Pickett, Ralph S, Maj WSO, Beaulaville, NC | 555 TFS | 432 TRW | 22 Dec 72 | 1.0 | MiG-21 | F-4D | Buick 01 | AIM-7 |

| Individual/Rank Crew position/Home town | USAF Sqn | Parent unit | Date | Credit | Type enemy aircraft | Type USAF aircraft | Radio callsign | Primary USAF weapon used |
|---|---|---|---|---|---|---|---|---|
| Priester, Durwood K, Maj<br>AC, Hampton, SC | 480 TFS | 366 TFW | 05 Jun 67 | 1.0 | MiG-21 | F-4C | Oakland 01 | 20mm |
| Rabeni, John J Jr, 1Lt<br>P, Southboro, Massachusetts | 480 TFS | 366 TFW | 05 Nov 66 | 1.0 | MiG-21 | F-4C | Opal 01 | AIM-7 |
| Radeker, Walter S III, Cpt<br>AC, Asheville, NC | 555 TFS | 8 TFW | 02 Jan 67 | 1.0 | MiG-21 | F-4C | Olds 04 | AIM-9 |
| Raspberry, Everett T Jr, Cpt<br>AC, Fort Walton Beach, Fla | 555 TFS | 8 TFW | 02 Jan 67<br>05 Jun 67 | 1.0<br>1.0 | MiG-21<br>MiG-17 | F-4C<br>F-4D | Ford 02<br>Drill 01 | AIM-9<br>AIM-7 |
| Retterbush, Gary L, Maj<br>AC, Lebanon, Indiana | 35 TFS | 388 TFW | 12 Sep 72<br>08 Oct 72 | 1.0<br>1.0 | MiG-21<br>MiG-21 | F-4E<br>F-4E | Finch 03<br>Lark 01 | 20mm<br>20mm |
| Richard, Lawrence G, Cpt (USMC)<br>AC, Lansdale, Pa | 58 TFS | 432 TRW | 12 Aug 72 | 1.0 | MiG-21 | F-4E | Dodge 01 | AIM-7 |
| Richter, Karl W, 1Lt<br>P, Holly, Michigan | 421 TFS | 388 TFW | 21 Sep 66 | 1.0 | MiG-17 | F-105D | Ford 03 | 20mm |
| Rilling, Robert G, Maj<br>P, South Berwick, Maine | 333 TFS | 355 TFW | 13 May 67 | 1.0 | MiG-17 | F-105D | Random 01 | AIM-9 |
| Ritchie, Richard S, Cpt<br>AC, Reidsville, NC | 555 TFS | 432 TRW | 10 May 72<br>31 May 72<br>08 Jul 72<br>08 Jul 72<br>28 Aug 72 | 1.0<br>1.0<br>1.0<br>1.0<br>1.0 | MiG-21<br>MiG-21<br>MiG-21<br>MiG-21<br>MiG-21 | F-4D<br>F-4D<br>F-4E<br>F-4E<br>F-4D | Oyster 03<br>Icebag 01<br>Paula 01<br>Paula 01<br>Buick 01 | AIM-7<br>AIM-7<br>AIM-7<br>AIM-7<br>AIM-7 |
| Roberts, Thomas S, Cpt<br>AC, LaGrange, Georgia | 45 TFS | 2 AD | 10 Jul 65 | 1.0 | MiG-17 | F-4C | Unknown 04 | AIM-9 |
| Roberts, William E Jr, 1Lt<br>P, Quitman, Oklahoma | 389 TFS | 366 TFW | 20 May 67 | 1.0 | MiG-21 | F-4C | Elgin 01 | AIM-9 |
| Rose, Douglas B, 1Lt<br>P, Chicago, Ill | 555 TFS | 8 TFW | 16 Sep 66 | 1.0 | MiG-17 | F-4C | Unknown 04 | AIM-9 |
| Rubus, Gary M, Cpt<br>AC, Banning, California | 307 TFS | 432 TRW | 15 Oct 72 | 1.0 | MiG-21 | F-4E | Buick 03 | 20mm |
| Russell, Donald M, Maj<br>P, Westbrook, Maine | 333 TFS | 355 TFW | 18 Oct 67 | 1.0 | MiG-17 | F-105D | Wildcat 04 | 20mm |
| Ryan, John D Jr, 1Lt<br>P, Pasadena, Texas | 13 TFS | 432 TRW | 17 Dec 67 | 1.0 | MiG-17 | F-4D | Gambit 03 | AIM-4 |
| Scott, Robert R, Col<br>P, Des Moines, Iowa | 333 TFS | 355 TFW | 26 Mar 67 | 1.0 | MiG-17 | F-105D | Leech 01 | 20mm |
| Sears, James F, 1Lt<br>P, Milan, Missouri | 389 TFS | 366 TFW | 26 Apr 67 | 1.0 | MiG-21 | F-4C | Cactus 01 | AIM-7 |
| Seaver, Maurice E Jr, Maj<br>P, Highland, California | 44 TFS | 388 TFW | 13 May 67 | 1.0 | MiG-17 | F-105D | Kimona 02 | 20mm |
| Sharp, Jerry K, 1Lt<br>P, Corpus Christi, Texas | 555 TFS | 8 TFW | 02 Jan 67 | 1.0 | MiG-21 | F-4C | Olds 02 | AIM-7 |
| Sheffler, Fred W, Cpt<br>AC, Akron, Ohio | 336 TFS | 8 TFW | 15 Aug 72 | 1.0 | MiG-21 | F-4E | Date 04 | AIM-7 |
| Shields, George I, 1Lt<br>WSO, Georgetown, Conn. | 469 TFS | 388 TFW | 12 Sep 72 | 1.0 | MiG-21 | F-4D | Robin 02 | AIM-9 |
| Sholders, Gary L, Cpt<br>AC, Lebanon, Oregon | 555 TFS | 432 TRW | 22 Dec 72 | 1.0 | MiG-21 | F-4D | Bucket 01 | Manoeuvring |
| Simmonds, Darrell D, Cpt<br>AC, Vernon, Texas | 435 TFS | 8 TFW | 06 Nov 67<br>06 Nov 67 | 1.0<br>1.0 | MiG-17<br>MiG-17 | F-4D<br>F-4D | Sapphire 01<br>Sapphire 01 | 20mm<br>20 mm |
| Simonet, Kenneth A, Maj<br>AC, Chicago, Illinois | 435 TFS | 8 TFW | 18 Jan 68 | 1.0 | MiG-17 | F-4D | Otter 01 | AIM-4 |
| Smallwood, John J, 1Lt<br>WSO, Atlanta, Georgia | 58 TFS | 432 TRW | 02 Jun 72 | 1.0 | MiG-19 | F-4E | Brenda 01 | 20mm |

| Individual/ Rank Crew position/ Home town | USAF Sqn | Parent unit | Date | Credit | Type enemy aircraft | Type USAF aircraft | Radio callsign | Primary USAF weapon used |
|---|---|---|---|---|---|---|---|---|
| Smith, Wayne O, 1Lt P, Clearwater, Florida | 435 TFS | 8 TFW | 18 Jan 68 | 1.0 | MiG-17 | F-4D | Otter 01 | AIM-4 |
| Smith, William T, 1Lt P, Wayne, Pennsylvania | 480 TFS | 35 TFW | 26 Apr 66 | 1.0 | MiG-21 | F-4C | Unknown 01 | AIM-9 |
| Squier, Clayton K, LtC AC, Oakland, California | 435 TFS | 8 TFW | 03 Jan 68 | 1.0 | MiG-17 | F-4D | Olds 01 | AIM-4 |
| Stone, John B, Cpt AC, Coffeeville, Miss | 433 TFS | 8 TFW | 02 Jan 67 | 1.0 | MiG-21 | F-4C | Rambler 01 | AIM-7 |
| Strasswimmer, Roger J, 1Lt P, Bronx, New York | 555 TFS | 8 TFW | 06 Jan 67 | 1.0 | MiG-21 | F-4C | Crab 02 | AIM-7 |
| Sumner, James M, 1Lt WSO, Manchester, Missouri | 35 TFS | 366 TFW | 23 May 72 | 1.0 | MiG-21 | F-4E | Balter 03 | 20mm |
| Suzanne, Jacques A, Cpt P, Lake Placid, New York | 333 TFS | 355 TFW | 12 May 67 | 1.0 | MiG-17 | F-105D | Crossbow 01 | 20mm |
| Swendner, William J, Cpt AC, Alamogordo, NM | 480 TFS | 35 TFW | 14 Jul 66 | 1.0 | MiG-21 | F-4C | Unknown 01 | AIM-9 |
| Taft, Gene E, LtC AC, Ventura, California | 4 TFS | 366 TFW | 29 Jul 72 | 1.0 | MiG-21 | F-4E | Pistol 01 | AIM-7 |
| Talley, James T, 1Lt P, Nixon, Texas | 480 TFS | 366 TFW | 14 May 67 | 1.0 | MiG-17 | F-4C | Speedo 03 | 20mm |
| Thies, Mack, 1Lt P, Houston, Texas | 390 TFS | 366 TFW | 01 May 67 | 1.0 | MiG-17 | F-4C | Stinger 01 | Manoeuvring |
| Thorsness, Leo K, Maj P, Las Vegas, Nevada | 357 TFS | 355 TFW | 19 Apr 67 | 1.0 | MiG-17 | F-105F | Kingfish 01 | 20mm |
| Tibbett, Calvin B, Cpt AC, Waynesville, Missouri | 555 TFS | 432 TRW | 09 Sep 72 16 Sep 72 | 1.0 1.0 | MiG-21 MiG-21 | F-4D F-4E | Olds 03 Chevy 03 | 20mm AIM-9 |
| Titus, Robert F, LtC AC, Hampton, Virginia | 389 TFS | 366 TFW | 20 May 67 22 May 67 22 May 67 | 1.0 1.0 1.0 | MiG-21 MiG-21 MiG-21 | F-4C F-4C F-4C | Elgin 03 Wander 01 Wander 01 | AIM-7 AIM-9 20mm |
| Tolman, Frederick G, Maj P, Portland, Maine | 354 TFS | 355 TFW | 19 Apr 67 | 1.0 | MiG-17 | F-105D | Nitro 03 | 20mm |
| Tracy, Fred L, Maj P, Goldsboro, NC | | 388 TFW | 29 Jun 66 | 1.0 | MiG-17 | F-105D | Unknown 02 | 20mm |
| Tuck, James E, Maj AC, Virgilina, Virginia | 480 TFS | 366 TFW | 05 Nov 66 | 1.0 | MiG-21 | F-4C | Opal 01 | AIM-7 |
| Turner, Samuel O, SSgt G, Atlanta, Georgia | | 307 SW | 18 Dec 72 | 1.0 | MiG-21 | B-52D | Brown III | 50-calibre |
| Vahue, Michael D, Cpt WSO, Battle Creek, Michigan | 523 TFS | 432 TRW | 16 Apr 72 | 1.0 | MiG-21 | F-4D | Papa 03 | AIM-7 |
| Voigt, Ted L II, 1Lt P, Nelsonville, Ohio | 555 TFS | 8 TFW | 14 Feb 68 | 1.0 | MiG-17 | F-4D | Nash 03 | 20mm |
| Volloy, Gerald R, Cpt WSO, Cincinnati, Ohio | 13 TFS | 432 TRW | 30 Mar 72 | 1.0 | MiG-21 | F-4D | Papa 01 | AIM-7 |
| Waldrop, David B III, 1Lt P, Nashville, Tennessee | 34 TFS | 388 TFW | 23 Aug 67 | 1.0 | MiG-17 | F-105D | Crossbow 03 | 20mm |
| Watson, George D, 1Lt WSO, Trenton, Missouri | 34 TFS | 388 TFW | 06 Oct 72 | 0.5 | MiG-19 | F-4E | Eagle 04 | Manoeuvring |
| Wayne, Stephen A, 1Lt P, Fairmount, Indiana | 433 TFS | 8 TFW | 13 May 67 20 May 67 | 1.0 1.0 | MiG-17 MiG-17 | F-4C F-4C | Harpoon 01 Tampa 03 | AIM-9 AIM-9 |
| Webb, Omri K III, 1Lt WSO, Leesville, SC | 34 TFS | 388 TFW | 05 Oct 72 | 1.0 | MiG-21 | F-4E | Robin 01 | AIM-7 |
| Wells, Norman E, 1Lt/Cpt P, Redwood City, California | 555 TFS | 8 TFW | 06 Jan 67 05 Jun 67 | 1.0 1.0 | MiG-21 MiG-17 | F-4C F-4C | Crab 01 Chicago 02 | AIM-7 AIM-9 |

| Individual/Rank Crew position/Home town | USAF Sqn | Parent unit | Date | Credit | Type enemy aircraft | Type USAF aircraft | Radio callsign | Primary USAF weapon used |
|---|---|---|---|---|---|---|---|---|
| Western, Robert W, 1Lt P, Carrollton, Alabama | 555 TFS | 8 TFW | 02 Jan 67 | 1.0 | MiG-21 | F-4C | Ford 02 | AIM-9 |
| Westphal, Curtis D, LtC AC, Bonduel, Wisconsin | 13 TFS | 432 TRW | 13 Oct 72 | 1.0 | MiG-21 | F-4D | Olds 01 | AIM-7 |
| Wetterhahn, Ralph F, 1Lt AC, New York City, NY | 555 TFS | 8 TFW | 02 Jan 67 | 1.0 | MiG-21 | F-4C | Olds 02 | AIM-7 |
| Wheeler, William H, Maj EWO, Fort Walton Beach, Fla | 357 TFS | 355 TFW | 19 Dec 67 | 1.0 | MiG-17 | F-105F | Otter 03 | 20mm |
| White, Sammy C, Cpt AC, Hot Springs, Arkansas | 4 TFS | 366 TFW | 19 Aug 72 | 1.0 | MiG-21 | F-4E | Pistol 03 | AIM-7 |
| Wiggins, Larry D, Cpt P, Houston, Texas | 469 TFS | 388 TFW | 03 Jun 67 | 1.0 | MiG-17 | F-105D | Hambone 03 | AIM-9/20mm |
| Williams, David O Jr, Col AC, Rockport, Texas | 435 TFS | 8 TFW | 14 Feb 68 | 1.0 | MiG-17 | F-4D | Killer 01 | AIM-7 |
| Wilson, Fred A Jr, 1Lt P, Mobile, Alabama | 333 TFS | 355 TFW | 21 Sep 66 | 1.0 | MiG-17 | F-105D | Vegas 02 | 20mm |
| Zimer, Milan, 1Lt P, Canton, Ohio | 389 TFS | 366 TFW | 20 May 67 22 May 67 22 May 67 | 1.0 1.0 1.0 | MiG-21 MiG-21 MiG-21 | F-4C F-4C F-4C | Elgin 03 Wander 01 Wander 01 | AIM-7 AIM-9 20mm |

Taken from *Aces and Aerial Victories, The USAF in South-east Asia, 1965-73*; Office of Air Force History.

# Appendix 5
# Aircraft and weapon combinations used in USAF Vietnam MiG victories

| USAF aircraft | Weapons/Tactics | MiG-17 | MiG-19 | MiG-21 | Total |
|---|---|---|---|---|---|
| F-4C | AIM-7 Sparrow | 4 | 0 | 10 | 14 |
| | AIM-9 Sidewinder | 12 | 0 | 10 | 22 |
| | 20mm gunfire | 3 | 0 | 1 | 4 |
| | Manoeuvring tactics | 2 | 0 | 0 | 2 |
| | | 21 | 0 | 21 | 42 |
| F-4D | AIM-4 Falcon | 4 | 0 | 1 | 5 |
| | AIM-7 Sparrow | 4 | 2 | 20 | 26 |
| | AIM-9 Sidewinder | 0 | 2 | 3 | 5 |
| | 20mm gunfire | 4 | 0 | 2 | 6 |
| | Manoeuvring tactics | 0 | 0 | 2 | 2 |
| | | 12 | 4 | 28 | 44 |
| F-4E | AIM-7 Sparrow | 0 | 2 | 8 | 10 |
| | AIM-9 Sidewinder | 0 | 0 | 4 | 4 |
| | AIM-9/20mm gunfire (combined) | 0 | 0 | 1 | 1 |
| | 20mm gunfire | 0 | 1 | 4 | 5 |
| | Manoeuvring tactics (2 F-4Es) | 0 | 1 | 0 | 1 |
| | | 0 | 4 | 17 | 21 |
| F-4D/F-105F | 20mm gunfire | 1 | 0 | 0 | 1 |
| | | 1 | 0 | 0 | 1 |
| F-105D | 20mm gunfire | 22 | 0 | 0 | 22 |
| | AIM-9 Sidewinder | 2 | 0 | 0 | 2 |
| | AIM-9/20mm gunfire (combined) | 1 | 0 | 0 | 1 |
| | | 25 | 0 | 0 | 25 |
| F-105F | 20mm gunfire | 2 | 0 | 0 | 2 |
| | | 2 | 0 | 0 | 2 |
| B-52D | 50-calibre gunfire | 0 | 0 | 2 | 2 |
| | | 0 | 0 | 2 | 2 |
| Grand totals | | 61 | 8 | 68 | 137 |